T0216366

Basic Concepts of Data and Error Analysis

Panayiotis Nicos Kaloyerou

Basic Concepts of Data and Error Analysis

With Introductions to Probability and Statistics and to Computer Methods

Springer

Panayiotis Nicos Kaloyerou
Department of Physics,
School of Natural Sciences
University of Zambia
Lusaka, Zambia

and

Wolfson College
Oxford, UK

ISBN 978-3-319-95875-0   ISBN 978-3-319-95876-7   (eBook)
https://doi.org/10.1007/978-3-319-95876-7

Library of Congress Control Number: 2018949632

© Springer Nature Switzerland AG 2018
This work is subject to copyright. All rights are reserved by the Publisher, whether the whole or part
of the material is concerned, specifically the rights of translation, reprinting, reuse of illustrations,
recitation, broadcasting, reproduction on microfilms or in any other physical way, and transmission
or information storage and retrieval, electronic adaptation, computer software, or by similar or dissimilar
methodology now known or hereafter developed.
The use of general descriptive names, registered names, trademarks, service marks, etc. in this
publication does not imply, even in the absence of a specific statement, that such names are exempt from
the relevant protective laws and regulations and therefore free for general use.
The publisher, the authors and the editors are safe to assume that the advice and information in this
book are believed to be true and accurate at the date of publication. Neither the publisher nor the
authors or the editors give a warranty, express or implied, with respect to the material contained herein or
for any errors or omissions that may have been made. The publisher remains neutral with regard to
jurisdictional claims in published maps and institutional affiliations.

This Springer imprint is published by the registered company Springer Nature Switzerland AG
The registered company address is: Gewerbestrasse 11, 6330 Cham, Switzerland

For my wife Gill and daughter Rose

Preface

This book began as a set of lecture notes for my laboratory sessions at the University of Swaziland many years ago and continued to be used as I moved to other universities. Over the years the lecture notes were refined and added to, eventually expanding into a short book. I also noticed that many books on practical physics were either advanced or very detailed or both. I felt, therefore, that there was a need for a book that focused on and highlighted the essential concepts and methods of data and error analysis that are of immediate relevance for students to properly write-up and present their experiments. I felt that my book filled this gap and so I decided it might be useful to publish it.

The original lecture notes comprised chapters one to four of the present book. However, the publisher suggested that the book should be extended, so it was decided to add chapter five, which is an introduction to probability and statistics, and chapter six, which introduces computer methods.

For students to get started with their experimental write-ups, only chapters one to four are needed. Though I have attempted to keep these chapters as simple as possible, slightly more advanced derivations of important formula have been included. Such derivations have been presented in separate sections so can be left out on first reading and returned to at a later time.

Chapter five aims to provide the theoretical background needed for a deeper understanding of the concepts and formula of data and error analysis. It is a stand-alone chapter introducing the basic concepts of probability and statistics. More generally, an understanding of the basic concepts of probability and statistics is essential for any science or engineering student. Though this chapter is a bit more advanced, I have tried to present it as simply and concisely as possible with the focus being on understanding the basics rather than on comprehensive detail. I have always felt that it helps understanding, and is, in any case, intrinsically interesting, to learn of the origins and the originators of the science and mathematics concepts that exist today. For this reason, I have included a brief history of the development of probability and statistics.

Chapter six introduces the computer methods needed to perform data and error calculations, as well as the means to represent data graphically. This is done by using four different computer software packages to solve two examples. However, it is hoped that the methods of this chapter will be used only after chapters one to four have been mastered.

I have added a number of tables and lists of mathematical formulae in the appendices for reference, as well as some biographies of scientists and mathematicians that have contributed to the development of probability and statistics.

It is hoped that this book will serve both as a reference to be carried to each laboratory session and also as a useful introduction to probability and statistics.

Lusaka, Zambia   Prof. Panayiotis Nicos Kaloyerou
June 2018

Acknowledgements

The author particularly thanks Dr. H. V. Mweene (Dean of Natural Sciences and member of the Physics Department, the University of Zambia) for meticulous proofreading of the manuscript and for numerous suggestions for improving the manuscript. Thanks are also due to Mr. E. M. Ngonga for careful proofreading.

Contents

About the Author

Prof. Panayiotis Nicos Kaloyerou was born on November 22, 1954 in Larnaka, Cyprus. He obtained his Ph.D. in Theoretical Physics in 1985 from Birkbeck College, London University, under the supervision of Prof. David Bohm, in which, based on Bohm's original idea, the Causal Interpretation of Boson Fields was developed. His first academic position was a Royal Society Fellowship held at the Institut Henri Poincaré (1987–1988). From 1990–1993, he held a Leverhulme Trust Research Fellowship at the University of Oxford. From 1993 to the present, he has worked mostly in physics departments in Africa: the University of Swaziland (1993–1996), the University of Botswana (1996–2001), and the University of Zambia (2005–present). His main publications include two Physics Reports articles: *An Ontological Basis for the Quantum Theory: a Causal Interpretation of Quantum Fields*, with D., Bohm and B. J. Hiley, (1987) and *The Causal Interpretation of the Electromagnetic Field* (1994).

List of Figures

Chapter 1
Units of Measurement

Chapters 1, 2 concern the treatment and presentation of experimental data. Error analysis begins in Chap. 3.

1.1 The Need for Units

When we give the results of a measurement, we need to give two pieces of information: one is the UNIT which we are using and the other is the NUMBER which gives the size, or 'magnitude', of the quantity when expressed in this unit.

For example, if you ask me how much money I have, I should say 'eight euros'; I mean that the amount of money I have is eight times the value of a one euro note. If I merely say 'eight' I might mean eight cents, or eight dollars, or eight cowrie-shells; the statement means nothing. We need the UNIT as well as the NUMBER.

Each quantity has its own proper unit. We measure length in metres, energy in joules and time in seconds.

1.2 Early Systems of Units

Before 1795, each country, or each part of a country, had its own units for length, mass and time. The **metric system** (metre-kilogram) was adopted in France in 1795. In this system, each unit is related to other units for the same quantity by multiples of ten, with Greek prefixes adopted for these multiples. This system has gradually been adopted all over the world, first for scientific work and later for trade and home use. In 1799 in France, the metre and kilogram were declared legal standards for all measurements.

© Springer International Publishing AG, part of Springer Nature 2018
P. N. Kaloyerou, *Basic Concepts of Data and Error Analysis*,
https://doi.org/10.1007/978-3-319-95876-7_1

British Imperial Units (foot-pound-second) were used in Great Britain since 1824 until the establishment in 1960 of the **International System of Units** (SI). Before 1960, many English speaking countries used British Imperial Units or units derived from them. An example is the system used in the United States, called the **United States Customary System**. In this system the pound (lb) is the unit of force and the slug is the unit of mass. Other British Imperial Units are the inch (in), foot (ft), yard (yd), mile, pound, stone, gallon (gal) (US and UK) and degrees Fahrenheit (°F). A number of these units are still in use today in English-speaking countries, especially in Great Britain and the United States, and can be found even in not-so-old textbooks.

During the 20th century the **CGS** (centimetre-gram-second), and **MKS** (metre-kilogram-second) units were also in common use before SI units were established in 1960.

1.3   The International System of Units. The SI System

In 1960 the 11th General Conference on Weights and Measures adopted the **Systeme International d'Unités** (metres-kilograms-seconds) or SI units for short. Before 1960, the SI system was called the MKS-system. The SI system has been almost universally adopted throughout the world. The SI system consists of the following basic units: length - metre (m), mass - kilogram (kg), time - second (s), electric current - ampere (A), temperature - kelvin (K),[1] quantity of substance - mole (m), and luminosity - candela (cd). The SI basic units, also called base SI units, are given in Table 1.1, and defined in Table 1.2. All other units are derived from these seven basic units (see Table 1.3).

1.4   Names and Symbols for SI Base Units

In printed material, slanting (italic) type is used for the 'symbol for quantity', and erect (Roman) type is used for the 'symbol for unit'. By 'symbol for quantity' we mean the symbol usually (but not necessarily always) used to represent a quantity; for example, energy is usually represented by the capital letter E, while the symbol for its unit, the joule, is J.

---

[1]The kelvin unit was named after Lord Kelvin, originally William Thompson (1824–1907). Lord Kelvin was born in Belfast, County Antrim, Ireland (now in Northern Ireland). He was an engineer, mathematician and physicist who made contributions mainly to thermodynamics and electricity. He obtained degrees from the universities of Glasgow and Cambridge and subsequently became Professor of Natural Philosophy (later named physics) at the University of Glasgow, a position he held for 53 years until he retired.

Table 1.1 Table of base SI units

Physical quantity	Symbol for quantity	SI base (Basic) unit	Symbol for unit
Length	l	metre	m
Mass	m	kilogram	kg
Time	s	second	s
Electric current	I	ampere	A
Temperature	T	kelvin	K
Amount of substance	n	mole	mol
Luminous intensity	I_v	candela	cd

Table 1.2 Table of definitions of base SI units

Unit	Definition of unit
Metre	The distance traveled by light in a vacuum in 1/299, 792, 458 of a second.
Kilogram	The mass equal to the mass of the international prototype kilogram of platinum-iridium kept at the International Bureau of Weights and Measures in Sévres, France.
Second	The duration of 9, 192, 631, 770 periods of the radiation corresponding to the transition between the hyperfine levels of the ground state of the cesium-133 atom.
Ampere	The current that, if maintained in two wires placed one metre apart in a vacuum, produces a force of 2×10^{-7} newtons per metre of length.
Kelvin	The fraction 1/273.16 of the thermodynamic temperature of the triple point of water.
Mole	The amount of substance containing as many elementary entities of the substance as there are atoms in 0.012 kg of carbon-12.
Candela	The intensity of radiation emitted perpendicular to a surface of 1/600,000 square metre of a blackbody at a temperature of freezing platinum at a pressure of 101,325 N/m^2.

Note that in the SI system we do not use capitals when writing the names of units, even when the unit name happens to be the name of a person. We write kelvin for the unit of temperature even though the unit is named after Lord Kelvin.

1.5 SI Derived Units

There is no base unit for, quantities such as speed. We have to build the unit for speed from two of the base units. A unit built in this way is called a 'derived unit'. Each derived unit is related in a very direct way to one or more base units.

Some derived units are given in Table 1.3. A more complete list is given in Appendix B.4. Notice that some, but not all of them, have special names of their own. Some units are named after the scientist who contributed most to the develop-

Table 1.3 Table of derived SI units

Physical quantity	Symbol for quantity	Name of unit	Symbol and/or definition in terms of base unit
Speed	v	Meter per second	$m \cdot s^{-1}$
Area	A	Meter-squared or square metre	m^2
Volume	V	Metre-cubed or cubic metre	m^3
Density	ρ	Kilogram per cubic metre	$kg \cdot m^{-3}$
Force	F	Newton	$N = m \cdot kg \cdot s^{-2}$
Energy	E	Joule	$J = m^2 \cdot kg \cdot s^{-2} = N \cdot m$

ment of the concept to which the unit refers, e.g., the newton (N), named after Sir Isaac Newton[2] and the joule (J) named after James Prescott Joule.[3]

1.6   Prefixes of 10

For many purposes, the SI base or derived unit is too small or too large. In the SI system we therefore use prefixes which make the unit larger or smaller. For example, the prefix 'kilo' means 'one thousand times', so a kilometre is one thousand times as long as a metre. The abbreviation for 'kilo' is 'k', so we can say 1 km = 1000 m. The prefixes with their abbreviations and meanings are tabulated in Table 1.4. A more complete list is given in Appendix B.5

---

[2]Sir Isaac Newton (1642–1727) was an English mathematician-natural philosopher (natural philosophy is now called physics) born in Lincolnshire, England. As well as mechanics, he made contributions in optics where he discovered that white light is composed of the colours of the rainbow by passing white light through a prism. His most significant contributions were to mechanics, where, building on fledgling concepts of motion he greatly advanced the subject using calculus for the first time, developed by himself and later, separately, by Leibnitz. He abstracted three aspects of motion and the production of motion as being fundamental and stated them as three laws, now called Newton's laws of motion. Based on these laws and on the introduction of his universal law of gravitation he performed many calculations among which was a quantitative description of the ebb and flow of tides, a description of many aspects of the motion of planets and of special features of the motion of the Moon and the Earth. He published this work in a remarkable book *Philosophiae Naturalis Principia Mathematica* (*Mathematical Principles of Natural Philosophy*) published in 1687. Another of his major works was his 1704 book, *Opticks*. Newton studied at the University of Cambridge and later became Lucasian professor of mathematics at the University of Cambridge.

[3]James Prescott Joule (1818–1889) was an English physicist born in Salford, England. He is most noted for demonstrating that heat is a form of energy, developing four methods of increasing accuracy to determine the mechanical equivalent of heat. He studied at the University of Manchester.

Table 1.4 Table of Prefixes of 10

Prefix	Symbol	Value
Tera	T	10^{12}
Giga	G	10^9
Mega	M	10^6
Kilo	k	10^3
Hecto	h	10^2
Deca	da	10^1
Deci	d	10^{-1}
Centi	c	10^{-2}
Milli	m	10^{-3}
Micro	μ	10^{-6}
Nano	n	10^{-9}
Pico	p	10^{-12}
Femto	f	10^{-15}
Atto	a	10^{-18}

1.7 Dimensional Analysis

It is often convenient to analyse formulae in terms of three basic quantities: mass, length and time. These are called dimensions (more generally we may introduce further 'dimensions' corresponding to other base SI units). They are given the special symbols

$$\text{mass, } M$$
$$\text{length, } L$$
$$\text{time, } T.$$

The main reason to analyse formulae in terms of dimension is to check that they are correct. This can best be explained by some examples.

Example 1.7.1 (Use of dimensional analysis to check the correctness of a formula (I).)
Suppose you are given the following formula:

$$\text{distance} = k \times \text{acceleration} \times \text{time}^2,$$

where k is a dimensionless constant. For this formula to be correct it must have the same dimensions on both sides.

Distance is a length and has dimension L
Time has dimension T
Acceleration has dimension LT^{-2}.

Substituting the dimensions into the formula we get

$$L = [LT^{-2}][T^2]$$
$$L = L,$$

which is clearly correct, since both sides of the formula are the same. We conclude that the formula is correct. Notice that dimensionless constants are ignored. Where physical constants have units, these are included in the dimensional analysis.

Example 1.7.2 (Use of dimensional analysis to check the correctness of a formula (II).)
Consider the following formula

$$V = \frac{P\pi a^3}{8\eta l}$$

Replace each quantity by its dimension as follows:

V = volume of liquid flowing per second = $\dfrac{\text{volume}}{\text{time}} = \dfrac{L^3}{T} = L^3T^{-1}$

P = pressure = $\dfrac{\text{force}}{\text{area}} = \dfrac{\text{mass} \times \text{accelaration}}{\text{area}} = \dfrac{MLT^{-2}}{L^2} = \dfrac{MT^{-2}}{L} = ML^{-1}T^{-2}$

π = 3.142 = dimensionless constant, hence ignored

8 = dimensionless constant, hence ignored

a = radius = L

η = viscosity = $\dfrac{\text{force} \times \text{time}}{\text{area}} = \dfrac{MLT^{-2} \times T}{L^2} = ML^{-1}T^{-1}$

l = length = L

Replace the physical quantities by their dimensions in the above formula.

$$L^3T^{-1} = \frac{[ML^{-1}T^{-2}][L^3]}{[ML^{-1}T^{-1}][L]}$$
$$L^3T^{-1} = T^{-1}L^2.$$

We conclude that the formula is incorrect, since the two sides of the formula are different. In fact, the correct formula is

$$V = \frac{P\pi a^4}{8\eta l}.$$

It can be checked thus,

$$L^3 T^{-1} = \frac{[ML^{-1}T^{-2}][L^4]}{[ML^{-1}T^{-1}][L]}$$
$$L^3 T^{-1} = T^{-1} L^3.$$

Since both sides are the same, we conclude that the formula is dimensionally correct. Note that this method can tell you when a formula is definitely wrong, but it cannot tell that a formula is definitely correct; it can only tell you that a formula is dimensionally correct.

1.8 Instruments for Length Measurement

Below are shown photographs of some widely used instruments for measuring dimensions of objects (outside dimensions, inside dimensions of hollow objects and depths). These are the Vernier caliper (Fig. 1.1), the micrometer screw gauge (Fig. 1.2) and the travelling microscope (Fig. 1.3). The Vernier caliper and the travelling microscope use an ingenious method for the accurate measurement of length dimensions, called the Vernier scale. Invented by Pierre Vernier[4] for the accurate measurement of fractions of millimetres. The micrometer screw uses another ingenious method for the accurate measurement of length dimensions based on the rotation of a screw system. Here, we will only describe the Vernier caliper and micrometer screw, since the travelling microscope uses the Vernier system for the accurate measurement of lengths.

1.8.1 The Principle of the Vernier Scale

The guide bar of the Vernier caliper shown in Figs. 1.1, 1.4, and 1.5 has two scales. The lower scale reads centimetres and millimetres, while the upper scale reads inches. Corresponding to this there are two scales on the Vernier, the lower one for measuring

---

[4]Pierre Vernier (1580–1637) was a French mathematician and government official born in Ornans, France. He was taught by his father, a scientist, and developed an early interest in measuring instruments. He described his new measuring instrument, now called the Vernier caliper, in his 1631 book *La Construction, l'usage, et les Propriétés du Quadrant Nouveau de Mathématiques* (*The Construction, Uses, and Properties of a New Mathematical Quadrant*).

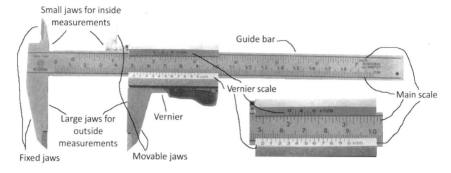

Fig. 1.1 Vernier calipers

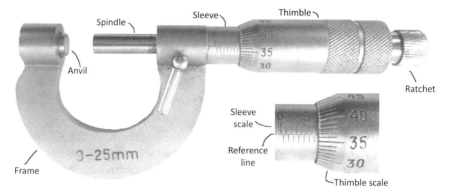

Fig. 1.2 Micrometer

fractions of millimetres, and the top one for measuring fractions of an inch. The two larger jaws are used to measure outer dimensions of an object, while the two smaller jaws are used to measure internal dimensions of an object. By sitting the guide bar on the bottom of an object and using the movable smaller jaw, depths can also be measured. We will first discuss the millimetre/centimetre scale.

The essence of understanding how the Vernier scale measures fractions of millimetres is to note that 20 divisions on the Vernier scale cover 39 mm of the main scale as shown in Fig. 1.4. The length of each division of the Vernier scale is therefore $39 \text{ mm}/20 = 1.95 \text{ mm}$. As a result, with the zeros of the two scales aligned as in Fig. 1.4, the first division of the Vernier scale is short of the 2 mm division of the main scale by 0.05 mm. If the Vernier scale is moved so that its first division is aligned with the 2 mm division of the main scale, the distance moved by the Vernier scale is 0.05 mm, and the main scale reads 0.05 mm. The reading on the main scale is indicated by the zero division of the Vernier scale. If the second division of the Vernier scale is aligned with the 4 mm division of the main scale the distance moved from the zero of the main scale is $2 \times 0.05 = 0.1 \text{ mm}$, so that the reading on the main scale is 0.1 mm. When the nth (where $0 \leq n \leq 20$) division of the Vernier scale is aligned

Fig. 1.3 Travelling microscope

Fig. 1.4 The figure shows 20 divisions of the Vernier scale covering 39 mm of the lower main scale, so that the divisions of the Vernier scale are each 1.95 mm long

with the $n \times 2$ mm division of the main scale, the Vernier has moved a distance of $n \times 0.05$ mm, corresponding to a reading of $n \times 0.05$ mm on the main scale. Clearly, for $n = 20$, the Vernier scale has moved a distance of $20 \times 0.05 = 1$ mm. We have taken the zero of the main scale as the reference for measuring the distance moved by the Vernier scale. However, it is clear that we can use any mm division of the main scale as a reference from which to take the distance moved by the Vernier scale. For example, take the 7 cm = 70 mm division of the main scale as the reference. If, say, the 9th division of the Vernier scale is aligned with a division on the main scale, then from our considerations earlier, we can say that the Vernier scale has moved a distance of $9 \times 0.05 = 0.45$ mm from the 70 mm division, and the corresponding reading on the main scale is $70 + 0.45 = 70.45$ mm $= 7.045$ cm. This then is how

Fig. 1.5 Reading Vernier calipers. The figure shows that the largest whole number of mm before the zero of the Vernier scale is 50 mm. The 7th division of the Vernier scale is aligned with a division of the main scale. The reading is therefore $50 + (7 \times 0.05) = 50.35$ mm or 5.035 cm

the Vernier scale measures fractions of millimetres. With this in mind, a reading on the Vernier scale is taken as follows:

The whole number of millimetres in a reading is taken as the largest number of millimetres on the main scale before the zero of the Vernier scale. The fraction of a millimetre is given by the division of the Vernier that is aligned with a division of the main scale. As an example, let us take the reading of the scales of the Vernier caliper shown in Fig. 1.5. The figure shows that the largest whole number of millimetres before the zero of the Vernier scale is 5 cm = 50 mm. The 7th division of the Vernier scale is aligned with a division of the main scale (which particular main scale division has no relevance). Thus, the reading is $50 + (7 \times 0.05) = 50.35$ mm. A simpler way to get the same result, is to read the Vernier scale as 0.35 mm, immediately giving a total reading of $50 + 0.35 = 50.35$ mm, as before. This simplified way of reading the Vernier scale is achieved by numbering every two divisions of the Vernier scale so that the 20 divisions are numbered from 0 to 10. The 0–10 numbering of the 20 Vernier scale divisions is shown in Figs. 1.1, 1.4, and 1.5. We may express the above length reading as a formula:

$$\text{length} = \text{MS} + (\text{VS} \times 0.05) \text{ mm}, \tag{1.1}$$

where MS is the main scale reading, and VS is the Vernier scale reading.

Some authors express the Vernier scale readings in terms of the least count (LC) of an instrument, defined as the smallest measurement that can be taken with an instrument, which is given by the formula

$$\text{LC} = \frac{\text{value of the smallest division on the main scale}}{\text{total number of divisions of the main scale}}.$$

For the Vernier calipers shown in Figs. 1.1, 1.4 and 1.5, the least count is LC = $\frac{1}{20} = 0.05$. We see that the LC is a dimensionless version of the 0.05 mm difference between 2 mm of the main scale and the 1.95 mm length of the Vernier scale divisions.

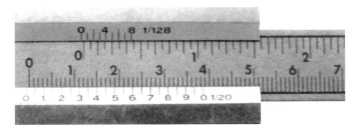

Fig. 1.6 Negative zero correction. Note that the zero of both the top (inches) and lower (millimetres/centimetres) Vernier scale is to the left of the zeros of the main scale. Since the 8th division on the Vernier scale is aligned with a division on the main scale, the negative zero correction to be added to the Vernier caliper reading is $(8 \times 0.05) = 0.4$ mm

A look at the top of the guide bar in Figs. 1.1, 1.4, and 1.5 shows a second inch scale with a smallest division of $\frac{1}{16}$ in. To read fractions of $\frac{1}{16}$ in, the Vernier has a second scale at its top. The principle is exactly the same as above. Here, 8 divisions of the top Vernier scale cover $\frac{7}{16}$ in of the top main scale, so that each division of the top Vernier scale has a length $\frac{7/16}{8} = \frac{7}{128}$ in. With the zeros of the two scales aligned the first top Vernier division is short of the $\frac{1}{16}$ in division of the main scale by $\frac{1}{128}$ in. The least count of the inch scale is LC $= \frac{1/16}{8} = \frac{1}{128}$ in. As an example let us take the reading from the top scales shown in Fig. 1.5. The main scale reading is taken to be the largest number of $\frac{1}{16}$ in before the Vernier zero, here MS $= \frac{31}{16}$ in $= 1\frac{15}{16}$ in. To find the fractional part of $\frac{1}{16}$ in, we can still use formula (1.1), but with 0.05 mm replaced by $\frac{1}{128}$ in. Noting that the closest Vernier division aligned with main scale division is the 5th Vernier division, and Using formula (1.1) with the $\frac{1}{16}$ in replacement, our length reading becomes $1\frac{15}{16} + 5 \times \frac{1}{128} = 1\frac{125}{128}$ in.

Finally, we need to consider the zero correction. When the caliper jaws are closed the reading should be zero. If it is not, readings must be corrected. When the zero of the Vernier is to the left of the zero of the main scale, as shown in Fig. 1.6, the zero correction is said to be negative. It is clear that all readings will be slightly low. Hence a negative zero correction must be added to the observed reading:

$$\boxed{\text{length} = \text{Vernier caliper reading} + \text{negative zero correction}}$$

When the zero of the Vernier is to the right of the zero of the main scale, as shown in Fig. 1.7, the zero correction is said to be positive. In this case, readings will be slightly high, so that a positive zero correction must be subtracted from the observed reading:

$$\boxed{\text{length} = \text{Vernier caliper reading} - \text{positive zero correction}}$$

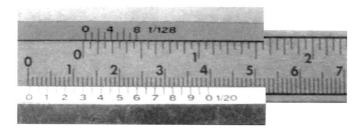

Fig. 1.7 Positive zero correction. Note that the zero of both the top (inches) and lower (millimetres/centimetres) Vernier scale is to the right of the zeros of the main scale. Since the 8th division on the Vernier scale is aligned with a division on the main scale, the positive zero correction to be subtracted from the Vernier caliper reading is $(8 \times 0.05) = 0.4\,\text{mm}$

1.8.2 The Micrometer

The principle of the micrometer is based on the fact that as a screw is turned by one revolution, it advances a distance equal to its pitch. The pitch of the screw is the distance between adjacent threads. When facing the ratchet, a clockwise rotation closes the micrometer, while an anticlockwise rotation opens it. Typically, a micrometer has a screw of pitch 0.5 mm and a thimble with 50 divisions. Thus, turning the thimble 1 division anticlockwise ($\frac{1}{50\text{th}}$ of a rotation) advances (opens) the micrometer spindle by $\frac{0.5}{50} = 0.01$ mm. The sleeve scale is marked in millimetres and half divisions of millimetres. Therefore, the sleeve division closest to the edge of the thimble gives either an integer millimetre reading or an integer plus a half integer millimetre reading, depending on whether the division is a millimetre or half millimetre division. Fractions of a half millimetre are read from the thimble. The thimble reading is the division on the thimble aligned with the line running along the sleeve, which we will call the reference line. For example, let us take the micrometer reading shown in Fig. 1.8. The nearest sleeve division before the edge of the thimble is the 7 mm division. The thimble division most closely aligned with the reference line is the 37th division. The micrometer reading is therefore $7 + (37 \times 0.01) = 7.37$ mm. As another example, consider reading the micrometer shown in Fig. 1.9 which involves a half millimetre sleeve division. Here, the nearest sleeve division before the thimble edge is the 6.5 mm division. The thimble division lies almost in the middle between the 23rd and 24th divisions, so we may take 23.5 as the closest division. The micrometer reading is therefore $6.5 + (23.5 \times 0.01) = 6.735$ mm.

As with the Vernier scale, we must consider the zero correction for the micrometer. For a fully closed micrometer the edge of the thimble should be exactly aligned with the zero of the sleeve when the zero of the thimble is aligned with the reference line.

If the zero of the thimble is above the reference line, as shown in Fig. 1.10, the zero correction is negative so that observed readings are too low. Again, a negative zero correction must be added to the observed reading:

Fig. 1.8 Reading a micrometer (I). The nearest sleeve division to the edge of the thimble is the 7 mm division. The thimble division most closely aligned with the reference line is the 37th division. The micrometer reading is therefore $7 + (37 \times 0.01) = 7.37$ mm

Fig. 1.9 Reading a micrometer (II). The nearest sleeve division to the edge of the thimble is the 6.5 mm division. The thimble division lies almost in the middle between the 23rd and 24th divisions, so we may take 23.5 as the closest division. The micrometer reading is therefore $6.5 + (23.5 \times 0.01) = 6.735$ mm

$$\boxed{\text{length} = \text{micrometer reading} + \text{negative zero correction}}$$

If the zero of the thimble is below the reference line, as shown in Fig. 1.11, the zero correction is positive and the observed readings are too high. Again, a positive zero correction must be subtracted from the observed reading:

$$\boxed{\text{length} = \text{micrometer reading} - \text{positive zero correction}}$$

For Vernier calipers it is easy to see why a zero correction is negative or positive, but for a micrometer a little more thought is needed. The two following examples address this issue.

Example 1.8.3 (Negative zero correction. The zero of the thimble is above the reference line.)
To see why a micrometer with the thimble zero above the reference line (negative zero correction) gives too low a reading, consider a micrometer with the zero above the reference line by 3 divisions, i.e., by $3 \times 0.01 = 0.03$ mm, as shown in Fig. 1.10. Also suppose that the length of an object has been measured and the observed reading is 6.1 mm. For the zero of the thimble to reach the reference line the spindle moves a

Fig. 1.10 Micrometer with a
negative zero correction. The
zero of the thimble is above
the reference line by 3
divisions so that the negative
zero correction to be added
to the micrometer reading is
$(3 \times 0.1) = 0.03$ mm

Fig. 1.11 Micrometer with a
positive zero correction. The
thimble zero is below the
reference line by 3 divisions
so that the positive zero
correction to be subtracted
from the micrometer reading
is $(3 \times 0.1) = 0.03$ mm

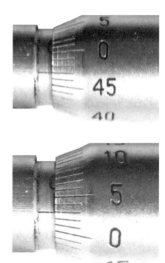

distance of 0.03 mm. To reach the observed reading of 6.1 mm, the spindle moves a
further distance of 6.1 mm. The correct length measurement of the object is the total
distance moved by the spindle, which is $0.03+6.1 = 6.13$ mm, clearly longer than the
observed reading. We see, then, why a micrometer with a negative zero correction
gives too low an observed reading, and, hence, why we must add a negative zero
correction: the distance traveled by the spindle for the thimble zero to reach the
reference line is not included in the observed reading.

Example 1.8.4 (Positive zero correction. The zero of the thimble is below the reference line.)
To see why a micrometer with the thimble zero below the reference line (positive
zero correction) gives too high a reading, consider a micrometer with the zero below
the reference line by 3 divisions, i.e., by $3 \times 0.01 = 0.03$ mm, as shown in Fig. 1.11.
Also suppose that the length of an object has been measured and the observed reading
is 6.1 mm. The zero of the thimble has a 'head start' of 3 divisions (or 0.03 mm) so
that to reach the reading of 6.1 mm the spindle only travels a distance of $6.1 - 0.03 =$
6.07 mm. Since the correct length measurement of the object is the distance moved
by the spindle, as stated above, the correct length of the object is 6.07 mm, which
is shorter than the observed reading. We see then why a micrometer with a positive
zero correction gives too high an observed reading and hence why we must subtract a
positive zero correction: The 3rd thimble division below the reference line indicates
a distance of 0.03 mm which the spindle has not moved, and this 'false' distance is
included in the observed reading, giving a value which is too high.

Chapter 2
Scientific Calculations, Significant Figures and Graphs

In a scientific experiment, we select observations which have to be made and we record them as accurately as the instruments we are using permit. Then we perform calculations on the data we have collected. Almost all our calculations are therefore carried out on measured quantities. A well-designed experiment, carefully carried out, is wasted if errors are made in calculation or if we claim more or less accuracy than we are entitled to do.

2.1 Accuracy of Measurements

All measurements are approximate. We make them by observing the position of something - a metre needle, a liquid level, the end of a string or spring - on some kind of scale. The scale is marked in divisions and the smallest division is usually chosen so that it can be easily seen by the eye, without using a lens. The last figure in our measurement is obtained by estimating fractions of this smallest division.

For example, suppose that we are using a good-quality plastic ruler to measure the length of an aluminium block. We place the zero of the scale exactly level with one side of the block and we look at the position of the other side of the block on the scale. We record the length of the block to the nearest millimetre and we find that it is just longer than 97 mm, but not as long as 98 mm. We estimate that six-tenths of this final millimetre should be included in the measurement. We can therefore say that the length of the block is 97.6 mm, correct to three significant figures. The first two of these figures can be regarded as 'certain' if the ruler has been graduated correctly. The third figure is an estimate; it could be wrong by 0.2 mm in either direction. This is usual for the final significant figure in a measurement.

Notice that the number of significant figures depends upon our method of observation. We cannot increase this number unless we use a different instrument. We can write the result down in a number of different ways but the number of significant

© Springer International Publishing AG, part of Springer Nature 2018

P. N. Kaloyerou, *Basic Concepts of Data and Error Analysis*,

https://doi.org/10.1007/978-3-319-95876-7_2

figures remains unchanged. For example, we can write the length of the block alternatively as 97.6 mm = 9.76 cm = 0.0976 m, but all lengths are given to **three significant figures**.

A good guide for the error in reading an instrument is as follows:

> The **Instrument Error Rule** states that the error in reading an instrument scale can be taken to be half the smallest division of the scale, i.e., ± a quarter of the smallest division (rounded down to the lowest single digit).

The requirement to round down to the lowest single digit is for consistency with rule 2.4.2 Sect. 2.4 for writing errors, namely, that an error should be given to 1 digit and be of the same power of 10 as the last (uncertain) significant figure of a measurement. For example, in the ruler measurement above the result is given as 97.6 mm. According to our rule, a ruler is accurate to half its smallest division, i.e., 0.5 mm, so that we should write the error in the answer as ± a half of 0.5 mm, i.e., 0.25 mm. Since there are two digits in the error 0.25 we must round down to the lowest single digit, namely, 0.2. Thus, the measurement together with error is written as 97.6 ± 0.2 mm.

2.2 Which Figures Are Significant?

In Sect. 2.1 above, we have used the phrase 'significant figure'. What is a significant figure? When are figures not significant?

Here are a few simple rules on significance, should any doubt ever arise.

1. All digits 1–9, and all zeros standing between such digits, are significant. So, for example, both these quantities are expressed to **four** significant figures:

$$4328 \ \text{m}$$
$$5.002 \ \text{mm}$$

2. Zeros written to the LEFT of the number and after a decimal point are used to express negative powers of ten, and are not considered significant as they can always be removed by putting the quantity into standard form or by changing the unit. For example, all these quantities are expressed to four significant figures.

$$0.4328 \ \text{km} = 4.328 \times 10^{-1} \ \text{km}$$
$$0.0004008 \ \text{mm} = 4.008 \times 10^{-4} \ \text{mm}$$
$$4008 \ \text{m} = 4.008 \ \text{km}$$

3. In a decimal number, that is, a number which includes a decimal point, all zeros written to the RIGHT of a non-zero digit are significant. For example:

> 2.011 has four significant figures
>
> 2.110 has four significant figures
>
> 30.000 has five significant figures

4. In a non-decimal number, that is, a number which does not include a decimal point, zeros written to the RIGHT of a non-zero digit are used to express powers of ten, and may or may not be significant. In such cases it is necessary to explicitly state the number of significant figures. For example, the number 5,359,783 stated to four significant figures reads 5,360,000. The four zeros are essential as they indicate powers of ten. Just by looking at this number we cannot tell the number of significant figures it represents. In such cases, it is essential to explicitly state the number of significant figures; for example, we should write 5,360,000 (3 sf), where the abbreviation 'sf' stands for 'significant figures'. Contrast this with the case of a decimal number (item 3 above) where the zeros to the right of the last non-zero digit indicate the number of significant figures.
5. In any case of doubt, it is good practice to state the number of significant figures being used. It is usual to abbreviate the statement like this:

$$\text{length} = 5.7\,\text{m} \ (2\,\text{sf})$$

2.3 Decimal Places

The difference between significant figures and decimal places must be clearly understood. For this reason and for completeness we consider what is meant by the number of decimal places. The abbreviation for 'decimal places' is 'dc. pl.'

1. The number of decimal places is the number of digits to the right of the decimal point. Consider the following examples where both the number of decimal places and the number of significant figures are given:

> 567.325 m (3 dc. pl.) (6 sf)
>
> 0.3254 m (4 dc. pl.) (4 sf)
>
> 23.32 m (2 dc. pl.) (4 sf)

2. Zeros between the decimal point and the last non-zero digit are counted. Zeros to the right of the last non-zero digit are only counted when the answer to a calculation or measurement is required to be specified to given number of decimal

places, otherwise they are ignored. In the following examples, the numbers are to be rounded to three decimal places:

$$567.30042 \, \text{m} \rightarrow 567.300 \, \text{m} \, (3 \, \text{dc. pl.}) \, (6 \, \text{sf})$$
$$2.35048 \, \text{m} \rightarrow 2.350 \, \text{m} \, (3 \, \text{dc. pl.}) \, (4 \, \text{sf})$$
$$0.03026 \, \text{m} \rightarrow 0.030 \, \text{m} \, (3 \, \text{dc. pl.}) \, (2 \, \text{sf})$$

Some examples where the number of decimal places is not specified:

$567.305000 \, \text{m} \rightarrow 567.305 \, \text{m} \, (3 \, \text{dc. pl.}) \, (6 \, \text{sf})$ - last three zeros are ignored.
$0.0054000 \, \text{m} \rightarrow 0.0054 \, \text{m} \, (4 \, \text{dc. pl.}) \, (2 \, \text{sf})$ - last three zeros are ignored.
$3.0003200 \, \text{m} \rightarrow 3.00032 \, \text{m} \, (5 \, \text{dc. pl.}) \, (6 \, \text{sf})$ - last two zeros are ignored.

2.4   Significant Figures and Experimental Uncertainty

The number of significant figures used for a measured quantity must not exceed the accuracy of the instrument used for measurement. Therefore, in measurement, the last (uncertain) significant figure should be to the same power of 10 as the error in the instrument. In Sect. 2.1 we saw that the error in measuring a length using a metre rule is ± 0.2 mm. Therefore, if we measure a length using a ruler and find 17.3 mm, we should write this measurement and its error as 17.3 ± 0.2 mm or 1.73 ± 0.02 cm. It would be pointless to try to estimate the length measurement to a greater accuracy, e.g., 17.28 mm, and it would be wrong to write the answer as 17.28 ± 0.2 mm.

Now, consider the case where measuring the thickness of a pencil using a micrometer yields 6.735 mm. The smallest division of a micrometer is 0.01 mm. We would therefore write the error in the measurement, according to the instrument error rule of Sect. 2.1, as ± 0.002 mm. The pencil thickness measurement is therefore written as 6.735 ± 0.002 mm.

We may abstract from the above examples the following two rules for the number of significant figures to be included in a measurement and the error in the measurement.

RULE 2.4.1	When writing the result of a measurement, include only one doubtful digit in the answer.

RULE 2.4.2	The error estimate should consist of only one digit, and should be of the same power of 10 as the last significant figure of the measurement.

The following are examples of acceptable ways to write a measurement together with its error:

$$g = 9.82 \pm 0.02 \, \text{m·s}^{-2} \, (3 \, \text{sf})$$
$$F = 10.2 \pm 0.1 \, \text{N} \, (3 \, \text{sf})$$
$$\omega = 3.237 \pm 0.002 \, \text{rad·s}^{-1} \, (3 \, \text{sf})$$

The following are examples of wrong ways to write a measurement together with its error:

$$g = 9.82 \pm 0.2 \, \text{m·s}^{-2} \, (3 \, \text{sf})$$
$$F = 10.2 \pm 1 \, \text{N} \, (3 \, \text{sf})$$
$$\omega = 3.237 \pm 0.02 \, \text{rad·s}^{-1} \, (3 \, \text{sf})$$

2.5 Calculations with Measured Quantities

When performing calculations with measured quantities certain definite rules should be followed:

RULE 2.5.1	Always convert your data to SI base or derived units before you start your calculation.

RULE 2.5.2	Never carry units through the calculation, but include the unit in the final result.

In multiplication or division, give your result to the number of significant figures of the **LEAST** precise of the data. For example, let us suppose that you wish to find the volume of a thick steel plate; and let it be supposed that the length, breadth, and thickness of the plate have all been measured using the same ruler:

$$\text{Length} = 255.4 \, \text{mm} = 0.2554 \, \text{m} = 2.554 \times 10^{-1} \, \text{m}$$
$$\text{Breadth} = 156.2 \, \text{mm} = 0.1562 \, \text{m} = 1.562 \times 10^{-1} \, \text{m}$$
$$\text{Thickness} = 5.6 \, \text{mm} = 0.0056 \, \text{m} = 5.6 \times 10^{-3} \, \text{m}$$

So Volume $= 2.554 \times 1.562 \times 5.6 \times 10^{-1} \times 10^{-1} \times 10^{-3} \times \text{m}^3$.

Multiplying using a calculator gives us

$$\text{Volume} = 2.2340349 \times 10^{-4} \, \text{m}^3$$

The result appears to be correct to EIGHT significant figures! Is this really so? No, it is not! The thickness is known to only two significant figures, and the second figure is doubtful; we might have chosen 5.7 mm, or 5.5 mm, because the last figure is an estimate. Calculate the volume again using these different thicknesses; you will find that only the FIRST figure remains unchanged! All that we can say for certain is that

$$\text{Volume} = 2.2 \times 10^{-4}\,\text{m}^3,$$

and even then the second figure is only an estimate. We see therefore that the answer after multiplication is only as accurate as the least accurate of the data.

Many calculations not only involve multiplications and divisions, but mixtures of additions, subtractions, multiplications and divisions in various combinations. Calculations may also include trigonometric, hyperbolic, logarithmic, exponential and other functions. A similar analysis to the multiplication example above will show that an answer to a calculation should be given to the same number of significant figures as the least accurate of the data. We may state this as a general rule:

RULE 2.5.3	The answer to a calculation should be given to the same number of significant figures as the least accurate of the data.

RULE 2.5.4	Always perform calculations to one more significant figure than you require in your answer. Then round off your final answer to the required number of significant figures. Always use the unrounded answer for subsequent calculations, if any.

For example, if a measuring instrument is accurate to three significant figures, then calculations should be performed to four significant figures. Your final answer will then be to four significant figures. Round off your answer to the required three significant figures. If you need to perform further calculations using the final answer, always use the unrounded answer, **NOT** the rounded answer. In our example, you would use the answer to four significant figures for subsequent calculation. Incidentally, this is general rule for performing calculations and should be used when solving text book problems, test or examination questions, or any other type of calculation.

2.5.1 Rules for Rounding Off

If the answer to a calculation is required to n sf, look at the $(n + 1)$th significant figure: if it is 4 or less leave the nth sf unchanged, if it is 5 or above then add 1 to the nth sf. Here are some examples:

$$84.632 \text{ rounded to 3 sf is } 84.6$$
$$84.672 \text{ rounded to 3 sf is } 84.7$$

84.696 rounded to 4 sf is 84.70
84.998 rounded to 4 sf is 85.00

Notice that a 1 is carried to left whenever a 9 has to be rounded up by 1, i.e., a 10 is added to the column to the left of the column in which the 9 appears.

The same rounding rule applies to decimal places:

84.632 rounded to 1 dc. pl.is 84.6
84.672 rounded to 1 dc. pl. is 84.7
84.696 rounded to 2 dc. pl. is 84.70
84.998 rounded to 2 dc. pl. is 85.00

2.6 Graph Drawing and Graph Gradients

This section provides general information on drawing graphs. Later, in Sect. 3.7 'Graphs', the points-in-pairs method for drawing the best straight line through a set of data points will be explained. The points-in-pairs method has the advantage that it provides a convenient method for finding the error in the slope of a straight-line graph.

2.6.1 Drawing and Using Graphs

We often present the data from a scientific experiment in the form of a **GRAPH** and then deduce a final result from the graph. Why do we do this? Because it makes it possible to see general trends in the data and because our final result then comes from all the data taken together, instead of from only part of it. Accuracy can therefore be much higher. Here are a set of rules for drawing and using graphs.

1. Use a sharp pencil.
2. Look at the data to be plotted. Notice whether quantities are increasing or decreasing in a reasonable manner. If any particular result seems wildly wrong, check the arithmetic; if this is correct, check the observation itself.
3. Sometimes the instructions for the experiment indicate which of your two variables is to be plotted on the horizontal x-axis and which is to be plotted on the vertical y-axis. The instruction 'plot A against B' means take B as the x variable and A as the y variable. If no instruction is given, plot the independent variable on the horizontal axis and the dependent variable on the vertical axis.
4. Leave wide margins all round your graph paper. Write a full title at the top of the graph. A typical title would be 'Velocity-Time Graph for a Trolley of Mass 0.750 kg When Accelerated by a Force of 0.30 N'.
5. Label each axis clearly with the quantity plotted and its unit.
6. Look at the **RANGE** of your data. If there are no negative values, only the positive portion of the corresponding axis need be drawn. Also, note whether

the origin or axes intercepts need to be included. If these are not needed, as is the case when only the slope needs to be calculated, the axes scales need not be started at zero, allowing the axes scales to be chosen so that the data points are well spread over the graph and not bunched in a small portion of the graph. For example, if the spread of data points is far from the origin, then including the origin will result in the data points being bunched in a small portion of the graph (see Fig. 3.8). We will return to this point at the end of Sect. 3.7.2.

7. Keeping the points made in item 6 in mind, choose your axis scales so that the data points are spread over the whole graph and that, if the graph is a straight line, the slope is not too close to either the x- or y-axis (see Fig. 2.1). Avoid using 1 cm to represent 3 or 7 or 1/3 or any other difficult or inconvenient number.

8. Mark your experimental points clearly. The following are examples of symbols which clearly mark experimental points:

$$\times, \ \otimes, \ \odot, \ \triangle \ or \ \triangledown \ \circ$$

A small square can also be used. It is important to show the data points clearly, so avoid using dots to mark them. These may be difficult to see and may even be obscured when the curve or line is drawn.

9. Errors in the dependent variable (the measured value) can be indicated with the symbol

,

where the length of the vertical line indicates the size of the error in the dependent variable. Sometimes, the following symbol is used to indicate errors in both the dependent and independent variables:

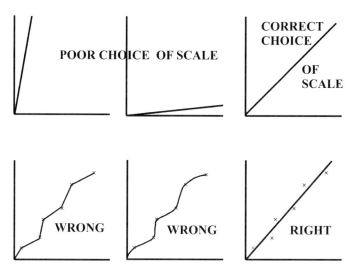

Fig. 2.1 Right and wrong choices of scale and connecting line

However, errors in the independent variable invariably contribute little to the overall error so that it is not generally useful to include error bars for the independent variable. Even for the dependent variable, it is not always useful to include error bars as they can complicate the graph. Very often, especially when computers are used to produce graphs of straight lines, including the data points is sufficient. The distance of the data point from the best straight line (to be discussed in later sections) itself gives a good visual indication of the error.

The following is a guide for when to use error bars and when not. Use error bars when either the error in each data point varies significantly, or more importantly, when the best straight line lies outside of the error range of many data points (ignore cases were only one or two data points lie beyond their error range from the best straight line since they are likely to result from wrong readings). It is not very useful to uses error bars when deviations of data points from the best straight line lie well within the error range of each data point.

10. If your best curve or line should pass through the origin, draw your best curve or line through the origin when judging the best straight line 'by eye'. But, do not do this when using the points-in-pairs method (to be described later) to draw the best straight line. In this method the distance from the origin (for lines that are supposed to pass through the origin) gives an estimate of the error.

11. Now check your graph. Do the data points crowd together or do they not make a recognisable line or curve? Then check the axes scales you chose. If these are okay, then check the experiment itself; did you use a reasonable range of values? Was the apparatus functioning correctly?

12. If your points are correctly plotted they will lie along, or close to, a recognizable straight line or a smooth curve. Decide which. If it is a curve, draw it in lightly with a single smooth movement of the pencil (a bendy ruler called a flexicurve helps to draw a smooth curve). If it is a straight line, draw in the 'best straight line' using a ruler. The line drawn must be straight; do not 'join the points' by short straight lines, or by short curves. Instead, you draw one smooth line, moving close to all the points, including the origin if this is one of your points, and leaving roughly as many points to one side as to the other (see Fig. 2.1). If one point is very far from the others, check it for arithmetical, experimental and plotting errors. If you cannot find what is wrong, leave the point marked clearly on your graph paper and put a bracket round it. Ignore it when drawing your line. Later, you will be shown the points-in-pairs method for drawing the best straight line.

13. If several graph lines are to be drawn on one set of axes, label the LINES clearly and separately by titles written close to the lines themselves. Use '×' to mark the data points for the first line, and some other symbol, such as '⊙', for the others.

14. A straight-line graph gives far more information than a curve does. If you know the relation between the variables then you can use the table in Sect. 3.7.4 to find what to plot to get a straight line.

2.6.2 Finding the Slope

The slope of a straight line is also called the gradient. A very steep line, nearly vertical, has a large slope. A gently-sloping line on the same axes has a small slope. A line which slopes upwards as we move from left to right has a positive slope; a line which slopes downward as we move from left to right has a negative slope.

The slope of a line is ALWAYS calculated 'in the units of the axes'. Thus, in the example below, the slope is in 'metres per second' or 'm/s' or 'ms^{-1}' and NOT in any other unit.

(a) *The Slope of a Straight-Line Graph*

To find the slope of a straight-line graph we construct a large right-angled triangle as shown in Fig. 2.2. *A* and *B* are points on the line; they are NOT plotted points.

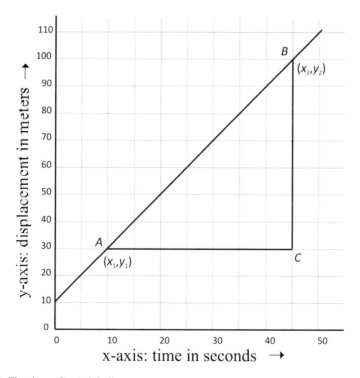

Fig. 2.2 The slope of a straight line

They are chosen so that lines AC and BC, drawn parallel to the two axes, are as far as possible convenient whole-number lines and not fractional ones.

The slope of the line is given by

$$\text{slope} = \frac{BC \text{ in units of the } y-\text{axis}}{AC \text{ in units of the } x-\text{axis}}$$

The distances AC and BC are found by CAREFUL reading of the co-ordinates (x_1, y_1) of point A, and the co-ordinates (x_2, y_2) of point B. Then

$$AC = (x_2 - x_1) = (45 - 10) \text{ s} = 35 \text{ s},$$

and

$$BC = (y_2 - y_1) = (100 - 30) \text{ m} = 70 \text{ m},$$

So the slope is

$$\text{slope} = \frac{70}{35} \text{ ms}^{-1} = 2 \text{ ms}^{-1}$$

Because this is a displacement-time graph, the slope is in units of velocity. If a slope is correctly calculated, then by carrying the units through the calculation, the correct units for the answer follow automatically.

When performing these slope calculations, always CHECK your values for the length of AC and BC by looking at the lines again; are your lengths correct? Students frequently make large errors in reading scales and co-ordinates.

(b) *The Slope of a Curve*

A curve, by definition, has a slope which varies along its length. We therefore have to find the slope of a curve at each point separately. For example, what is the slope of the curve at point P in Fig. 2.3? To find out, we draw a TANGENT at P and then determine the slope of the tangent. The slope at P is positive. If we repeat the process at Q, we shall find a negative slope. Notice that at points R and S the slope is zero.

Slope at Point P

At point P, the slope of the curve can be found by determining the slope of the tangent AB:

$$\text{slope of } AB = \frac{BC}{AC} = \frac{y_2 - y_1}{x_2 - x_1}$$
$$= \frac{55 - 15}{24 - 4}$$
$$= \underline{2 \text{ ms}^{-1}}$$

This slope is positive; the line slopes upwards to the right.

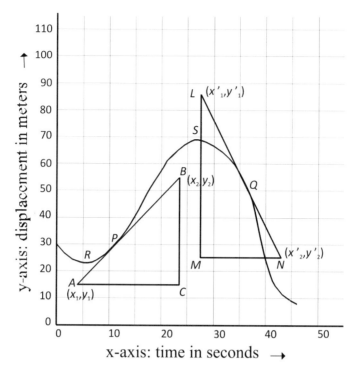

Fig. 2.3 The slope of a curve

Slope at Point Q

At point Q, the slope of the curve can be obtained by finding the slope of the tangent LN.

$$\text{slope of } LN = \frac{LM}{NM} = \frac{y_2' - y_1'}{x_2' - x_1'}$$

$$= \frac{25 - 86}{46 - 27.5}$$

$$= \underline{-3.3\,\text{ms}^{-1}}$$

This slope is negative; the line slopes downwards to the right. At points R and S, any tangent drawn would be parallel to the x-axis and any triangle drawn would have zero height. The slope at these points is therefore zero.

Chapter 3
Error Analysis

3.1 Random Errors

REPEATED READINGS - Never be satisfied with a single reading; repeat the measurement. This procedure will improve the precision of results; it can also show up careless mistakes.

RANDOM ERRORS - All measurements are subject to random errors and these spread the readings about the true values. Sometimes the reading is too high, sometimes too low. With repeated readings, random errors tend to cancel out.

If n readings are taken then the best estimate is the mean (or average):

$$\boxed{\text{MEAN}} \quad \boxed{\bar{x} = \frac{x_1 + x_2 + x_3 + \cdots + x_n}{n}}$$

Example 3.1.1 (Calculation of the mean) Suppose you are measuring the volume of water flowing through a tube in a given time. Five readings of this quantity may yield the values:

$$436.5, \ 437.3, \ 435.9, \ 436.4, \ 436.9 \ \text{cm}^3$$

(If one reading differs too much from the others, it may be due to a mistake. Since a bad reading will influence the average, it may be better to neglect it.)
The mean is given by

$$\bar{x} = \frac{436.5 + 437.3 + 435.9 + 436.4 + 436.9}{5} \ \text{cm}^3$$

$$= 436.6 \ \text{cm}^3$$

© Springer International Publishing AG, part of Springer Nature 2018
P. N. Kaloyerou, *Basic Concepts of Data and Error Analysis*,
https://doi.org/10.1007/978-3-319-95876-7_3

The more readings you take, the more reliable your results.

Example 3.1.2 (Two experiments showing a different spread of results)

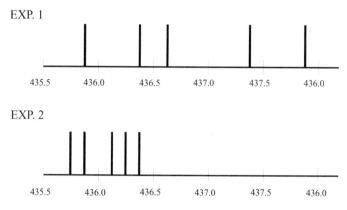

Fig. 3.1 Results from two different experiments showing different spreads

The results of experiment 2 are regarded as more reliable because of the smaller spread.

Generally, a smaller spread indicates a more reliable experiment. The STANDARD DEVIATION is a means of measuring the spread of results (Fig. 3.1). To calculate the standard deviation:

FIRST: Calculate the mean $\bar{x}$

SECOND: Calculate the <u>residuals</u>, $d_1, d_2, \ldots, d_n$. The residuals are the differences between the individual readings and the mean:

RESIDUALS	$d_1 = x_1 - \bar{x}$ $d_2 = x_2 - \bar{x}$ $\vdots \quad \vdots$ $d_n = x_n - \bar{x}$

Then, the STANDARD DEVIATION, σ, is defined by

$$
\boxed{\text{STANDARD DEVIATION}} \quad \boxed{\sigma = \left(\frac{d_1^2 + d_2^2 + d_3^2 + \cdots d_n^2}{n} \right)^{\frac{1}{2}}} \qquad (3.1)
$$

The above formula for standard deviation is precise when all possible values of a given quantity are available. The entirety of possible values is called the POPULATION and we may call the above formula the POPULATION STANDARD DEVIATION. Often, as in nearly all experiments, only a portion of all possible values, called a SAMPLE, is available. When only a sample is available, the best estimate of the population standard deviation is the SAMPLE STANDARD DEVIATION, σ_s, given by

$$
\boxed{\text{SAMPLE STANDARD DEVIATION}} \quad \boxed{\sigma_s = \left(\frac{d_1^2 + d_2^2 + d_3^2 + \cdots d_n^2}{n-1} \right)^{\frac{1}{2}}} \qquad (3.2)
$$

In what follows, we will use the term 'standard deviation' to mean 'population standard deviation', but we will use the full name when we want to refer to the 'sample standard deviation'.

Example 3.1.3 (Calculation of standard deviation) Calculate the standard deviation of the volume readings in Example 3.1.1 (i.e., 436.5, 437.3, 435.9, 436.4 and 436.9 cm^3).

Solution

$$\bar{x} = 436.6 \text{ cm}^3 \quad \text{(from Example 3.1.1)}$$

$$d_1 = 436.5 - 436.6 = -0.1$$
$$d_2 = 437.3 - 436.6 = 0.7$$
$$d_3 = 435.9 - 436.6 = -0.7$$
$$d_4 = 436.4 - 436.6 = -0.2$$
$$d_5 = 436.9 - 436.6 = 0.3$$

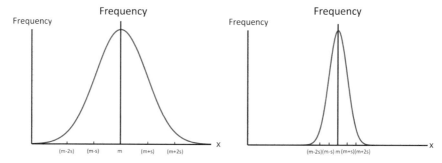

Fig. 3.2 Example of frequency graphs

$$\sigma = \left(\frac{d_1^2 + \cdots d_5^2}{n}\right)^{\frac{1}{2}} = \left(\frac{0.01 + 0.49 + 0.49 + 0.04 + 0.09}{5}\right)^{\frac{1}{2}}$$

$$\sigma = \left(\frac{1.12}{5}\right)^{\frac{1}{2}} = 0.224^{\frac{1}{2}}$$

$$\sigma = 0.47 \text{ cm}^3$$

$$\sigma \approx 0.5 \text{ cm}^3$$

If we take a large number of readings, we may plot a graph of the distribution of the readings, i.e., the number of occurrences of a given value of a variable (called the frequency) versus the value of the variable. Such graphs are called frequency graphs and the curves frequency curves.

The second graph in Fig. 3.2 shows readings that are more closely spaced about the mean. The smaller spread is indicated by a smaller standard deviation. In fact, the standard deviation is a measure of the spread of a set of readings.

Knowing the standard deviation is not enough; we need some way of estimating how far the mean is from the true value. The problem is that although the true value exists, we cannot know what this value is.

A measure of how far the mean is from the true value (the error in the mean), derived from statistical arguments, a derivation that we will give in the next subsection, is the STANDARD ERROR IN THE MEAN, s_m, defined by

$$\boxed{\begin{array}{c} \text{STANDARD ERROR} \\ \text{IN THE MEAN} \end{array} \quad s_m = \frac{\sigma}{(n-1)^{\frac{1}{2}}}} \qquad (3.3)$$

It is such that the true value x has a 68% chance of lying within $\pm s_m$ of the mean value and a 95% chance of lying within $\pm 2s_m$, etc. Thus, s_m is the required measure of how close the mean value of the given sample, $\bar{x}$, is to the unknown true value.

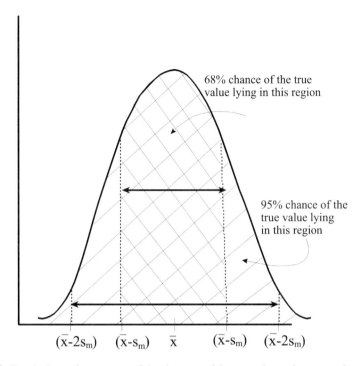

Fig. 3.3 Standard error is a measure of the closeness of the true value to the mean value

We may represent the chance of the true value being found near the mean graphically. This is shown in Fig. 3.3.

Example 3.1.4 (Calculation of standard error) Calculate the standard error for the five readings of Example 3.1.1.

Solution

$$\text{mean} = \bar{x} = 436.6 \text{ cm}^3 \text{ (from Example 3.1.1.)}$$

$$\text{Standard deviation} = \sigma = 0.47 \text{ cm}^3 \text{ (from Example 3.1.3)}$$

$$s_m = \frac{\sigma}{(n-1)^{\frac{1}{2}}} = \frac{0.47}{(5-1)^{\frac{1}{2}}} = \frac{0.47}{2}$$
$$= 0.235 \text{ cm}^3 \approx 0.2 \text{ cm}^3$$

The final answer is written as

$$\bar{x} = (436.6 \pm 0.2) \text{ cm}^3.$$

Notice that s_m depends both on the standard deviation σ and on the number n of readings.

3.1.1 Derivation of the Formula for the Standard Error in the Mean

Consider n successive measurements of some quantity yielding the values

$$x_1, x_2, \ldots, x_n.$$

The number n of measurements need not be large. In student laboratory experiments, the number of measurements is typically between 10–20. In research experiments, the number may well be much higher. As stated earlier, the best estimate of the value of the measured quantity is the mean of these n measurements

$$\bar{x}_k = \left[\frac{1}{n} \sum_{i=1}^{n} x_i \right]_k. \tag{3.4}$$

The error e_i in the ith measurement x_i is

$$e_i = x_i - X, \tag{3.5}$$

where X is the impossible to know true value of the quantity.

The error in the mean is given by

$$E_k = \bar{x} - X = \frac{1}{n} \sum_{i=1}^{n} x_i - X = \frac{1}{n} \sum_{i=1}^{n} x_i - \frac{nX}{n} = \frac{1}{n} \sum_{i=1}^{n} (x_i - X) = \frac{1}{n} \sum_{i=1}^{n} e_i. \tag{3.6}$$

From Eq. (3.6) we get

$$E_k^2 = \frac{1}{n^2} \sum_{i=1}^{n} \sum_{j=1}^{n} e_i e_j = \frac{1}{n^2} \sum_{i=1}^{n} e_i^2 + \frac{1}{n^2} \sum_{i=1}^{n} \sum_{j=1, j \neq i}^{n} e_i e_j = \frac{1}{n} \langle e^2 \rangle + \frac{1}{n^2} \sum_{i=1}^{n} \sum_{j=1, j \neq i}^{n} e_i e_j. \tag{3.7}$$

Now consider a large number N of sets of n measurements. We label each set of n measurements by $k = 1$ to N. Each set has its own mean $\bar{x}_k$, its own standard deviation σ_k of the errors in each measurement x_i, and its own error E_k in the mean. The mean of the kth set of n measurements is given by Eq. (3.4).

The square of the standard deviation (variance), σ_k^2, of the errors in the x_i measurements of the kth set of n measurements is given, using formula (3.5), by

$$\sigma_k^2 = \left[\frac{1}{n}\sum_{i=1}^{n}(x_i - X)^2\right]_k = \left[\frac{1}{n}\sum_{i=1}^{n}e_i^2\right]_k = \langle e^2 \rangle_k. \tag{3.8}$$

The square of the standard deviation (variance) σ_t^2 of the errors in each measurement of the grand distribution formed by lumping together the measurements of the N sets is given by

$$\sigma_t^2 = \frac{1}{N}\sum_{k=1}^{N}\sigma_k^2 = \frac{1}{N}\sum_{k=1}^{N}\langle e^2 \rangle_k = \langle e^2 \rangle. \tag{3.9}$$

The error in the mean of the kth set of n measurements is given by Eq. (3.6).

Now, the standard deviation s_m of the error in the means of the N sets of n measurements is the standard error in the mean that we are seeking. Its square, the standard variance of the mean, is given by

$$s_m^2 = \frac{1}{N}\sum_{k=1}^{N}(\bar{x}_k - X)^2 = \frac{1}{N}\sum_{k=1}^{N}E_k^2 = \langle E^2 \rangle. \tag{3.10}$$

Next, we want to establish a relation between $\langle E^2 \rangle$ and $\langle e^2 \rangle$. Substituting Eq. (3.7) for each E_k in Eq. (3.10) gives

$$s_m^2 = \frac{1}{N}\sum_{k=1}^{N}\left[\frac{1}{n}\langle e^2 \rangle + \frac{1}{n^2}\sum_{i=1}^{n}\sum_{j=1,j\neq i}^{n}e_ie_j\right]_k$$

$$= \frac{1}{N}\sum_{k=1}^{N}\frac{1}{n}\langle e^2 \rangle_k + \frac{1}{N}\sum_{k=1}^{N}\left[\frac{1}{n^2}\sum_{i=1}^{n}\sum_{j=1,j\neq i}^{n}e_ie_j\right]_k.$$

Since e_i and e_j are independent, negative terms cancel positive terms in the double sum, so that the double sum term equals zero, giving

$$s_m^2 = \frac{1}{N}\sum_{k=1}^{N}\frac{1}{n}\langle e_k^2 \rangle = \frac{1}{n}\langle e^2 \rangle. \tag{3.11}$$

Equations (3.10) and (3.11) establish the relation we want:

$$\langle E^2 \rangle = \frac{1}{n} \langle e^2 \rangle. \tag{3.12}$$

Substituting Eqs. (3.9) and (3.10) into Eq. (3.12) and taking the square root gives

$$s_m = \frac{\sigma_t}{\sqrt{n}}. \tag{3.13}$$

We do not yet have a formula for the standard error in the mean since σ_t is an unknown quantity. We therefore need a way to approximate this term. We can do this as follows: From Eqs. (3.5) and (3.6) we get

$$x_i - \bar{x} = e_i - E_k. \tag{3.14}$$

Substituting this into the definition of standard deviation we get

$$\sigma'^2 = \frac{1}{n} \sum_{i=1}^{n} (x_i - \bar{x})^2 = \frac{1}{n} \sum_{i}^{n} (e_i - E_k)^2$$

$$= \frac{1}{n} \sum_{i=1}^{n} \sum_{j=1}^{n} e_i e_j - 2 E_k \frac{1}{n} \sum_{i=1}^{n} e_i + E_k^2$$

$$= \frac{1}{n} \sum_{i=1}^{n} e_i^2 + \frac{1}{n} \sum_{i=1}^{n} \sum_{j=1, j \neq i}^{n} e_i e_j - 2 E_k \frac{1}{n} \sum_{i=1}^{n} e_i + E_k^2$$

$$= \frac{1}{n} \sum_{i=1}^{n} e_i^2 + \frac{1}{n} \sum_{i=1}^{n} \sum_{j=1, j \neq i}^{n} e_i e_j - 2 E_k^2 + E_k^2,$$

where we have used Eq. (3.6). The double sum gives zero since the negative terms cancel the positive terms pairwise. Hence

$$\sigma'^2 = \frac{1}{n} \sum_{i=1}^{n} e_i^2 - E_k^2.$$

This relation is for one of the N sets of n measurements. Summing over all N we get

$$\langle \sigma^2 \rangle = \frac{1}{N} \sum_{k=1}^{N} \sigma_k'^2 = \frac{1}{N} \sum_{k=1}^{N} \left[\frac{1}{n} \sum_{i=1}^{n} e_i^2 \right]_k - \frac{1}{N} \sum_{i=1}^{N} E_k^2.$$

Using Eqs. (3.8), (3.9) and (3.10) we get

$$\langle \sigma^2 \rangle = \frac{1}{N} \sum_{k=1}^{N} \langle e^2 \rangle_k - \langle E^2 \rangle = \langle e^2 \rangle - \langle E^2 \rangle.$$

Substituting Eqs. (3.9) and (3.10) into the above equation gives

$$\langle \sigma^2 \rangle = \sigma_t^2 - s_m^2.$$

Substituting Eq. (3.13) gives

$$\langle \sigma^2 \rangle = n s_m^2 - s_m^2.$$

Rearranging and taking the square root gives:

$$s_m = \frac{\sqrt{\langle \sigma^2 \rangle}}{(n-1)^{\frac{1}{2}}}.$$

Now $\langle \sigma^2 \rangle$ is still an unknown quantity. But, a very good estimate of it is the square of the standard deviation σ^2 of the set of the n actual measurements, i.e., $\langle \sigma^2 \rangle \approx \sigma_k^2$, giving

$$s_m = \frac{\sigma}{(n-1)^{\frac{1}{2}}},$$

where we have dropped the k subscript. This completes the derivation of the formula for the standard error in the mean.

3.2 Systematic Errors

A random error spreads results about the true value and from the equation for s_m it is clear that by taking a large enough number of readings n, the random error can be made arbitrarily small. However, there are other errors, called systematic errors. With systemic errors the readings are not spread about the true value, but about some displaced value. Systematic errors will therefore cause the mean to be shifted away from the true value. In this case, simply repeating the measurements will not reduce the systematic errors.

It is customary to distinguish between an ACCURATE and a PRECISE measurement. An ACCURATE measurement is one in which systematic errors are small; a PRECISE measurement is one in which the random errors are small. Some examples of systematic errors are:

1. Parallax errors (these can occur when reading a pointer on a scale).
2. Zero errors (these occur when an instrument is not properly set at zero).
3. Inaccurate instrument scales (these are errors inherent in instrument scales, such as those of rulers or micrometers).

4. Inaccurate times from a stop-watch that is too slow or too fast.
5. Inaccurate mass readings when standard masses are too light or too heavy.

All that you can do with systematic errors is to estimate them (i.e., take a guess, guided by an understanding of the physical processes involved). For example, recall the Instrument Error Rule of Sect. 2.1, which states that the accuracy of an instrument scale can be taken to be half of the smallest division, i.e., ± a quarter of the smallest division.

Example 3.2.1 (Instrument error rule) For a length of, say, 15.6 mm measured by an ordinary ruler, the error estimate would be ±0.2 (see Sect. 2.1) and we would write

$$L = (15.6 \pm 0.2) \text{ mm}.$$

3.3 Combining Systematic and Random Errors

Consider again Example 3.1.1 where five readings of the volume of water flowing through a tube are taken. So far we have found $\bar{x} = 436.6$ cm^3 and $s_m = 0.2$ cm^3. Now consider what systematic errors may have occurred in taking the readings:

1. The person timing may have had a tendency to start or stop the stop-watch too soon or too late. Let us suppose the person had a tendency to start the stop-watch too soon or too late by 1/5 s, so that time readings are either longer or shorter than the true values. This is a common type of systematic error and arises because individuals respond to aural and visual stimuli in different ways. Now, suppose further that the volume of the flowing water was measured in a time interval of 4 min = 240 s, then

$$\text{volume flowing in one second} = \frac{436.6}{240} \text{ cm}^3$$

$$\text{volume flowing in } \frac{1}{5} \text{ one second} = \frac{436.6}{240} \times \frac{1}{5} \text{ cm}^3$$

$$= 0.36 \approx 0.4 \text{ cm}^3.$$

Since the stop watch was started too late or too soon by 1/5 s, the systematic error in measuring the volume flowing in 240 s is the volume flowing in 1/5 s, i.e., 0.4 cm^3. Thus, the error in the volume flowing in 240 s is:

$$\text{Systematic error} = 0.4 \text{ cm}^3.$$

Fig. 3.4 The curved liquid
surface can give rise to
systematic parallax errors

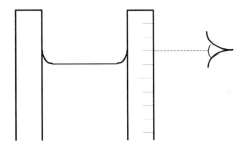

2. Because the surface of a liquid is curved, systematic parallax errors may occur in reading the volume of the water. To account for this, we estimate a further error of 0.2 cm^3 in measuring the volume of the water (Fig. 3.4).

We now have three errors in measuring the volume: The standard error $s_m = 0.2$ cm^3 (calculated in Example 3.1.4), and the two systematic errors of 0.4 cm^3 and 0.2 cm^3 calculated above. The question is how to combine these errors to get the total error. We could simply add them to get

$$0.2 + 0.4 + 0.2 = 0.8 \text{cm}^3.$$

This would be correct if all the errors tended to push the value in the same direction - either up or down. In some cases this may occur. Generally, however, summing the errors would be an over-estimate. Since some errors may push the value down, whilst others push the value up, they have a tendency to cancel each other, reducing the overall error.

From detailed statistical arguments, it is found that the best estimate for combining random and systematic errors is given by

$$\Delta E = [(\Delta e_1)^2 + (e_2)^2 + (e_3)^2 + \ldots + (e_n)^2]^{\frac{1}{2}}$$

$\Delta E = $ total error, where the Δe_i may be standard or systematic errors.

Thus, the total error for our example is

$$\Delta E = (0.2^2 + 0.4^2 + 0.2^2)^{\frac{1}{2}}$$
$$= (0.04 + 0.16 + 0.04)^{\frac{1}{2}}$$
$$= 0.24^{\frac{1}{2}}$$
$$= 0.5 \text{ cm}^3$$

We will derive the formula for ΔE in subsection 3.4.3.

3.4   Errors in Formulae

In this section we look at how errors in measured quantities (determined as described in the previous section) produce errors in calculated quantities (from formulae having various forms). How errors in measured quantities produce errors in calculated quantities is called the *propagation of errors*. The calculated quantity, which from here on we shall label as Z, may depend on one measured quantity A, or it may depend on two or more measured quantities A, B, ... When Z depends on only one quantity A we want to derive formulae that express the error ΔZ in Z in terms of the error ΔA in A. For a Z which depends on more than one measured quantity we want to determine ΔZ in terms of ΔA, ΔB, . . .; in other words, we want to derive formulae for combining errors. We will restrict our derivations to one and two-quantity formulae, and merely state the error formulae for more than two measured quantities. In some cases, it is more convenient to express the relative error $\frac{\Delta Z}{Z}$ instead of the error ΔZ. In what follows, measured quantities will be labeled by either A, B, C, . . . or by A_1, A_2, A_3, . . . for general formula, while k, l and n are constants.

Example 3.4.2 (Propagation of errors for a one-quantity formula) We have measured the radius r of a wire with error Δr and want to calculate the cross-sectional area A from the formula

$$A = \pi r^2. \tag{3.15}$$

One approach is the simple direct approach where $r + \Delta r$ and $A + \Delta A$ is substituted into Eq. (3.15). We get

$$A + \Delta A = \pi [r + \Delta r]^2 = \pi (r^2 + 2r \Delta r + (\Delta r)^2]$$

For a good experiment Δr is small, so that $(\Delta r)^2$ can be neglected to a good accuracy. Even so, the expression is still cumbersome. Moreover, more detailed statistical arguments show that it is too large an error estimate.

Example 3.4.3 (Propagation of errors for multi-quantity formula) By measuring the length l, width w and height h of a water tank with errors Δl, Δw Δh, respectively, we want to calculate its volume from the formula

$$V = lwh.$$

Again, by direct substitution, we get an error estimate

$$V + \Delta V = (l + \Delta l)(w + \Delta w)(h + \Delta h)$$

This is even more cumbersome than the previous example, and, again, more detailed statistical arguments show that this approach gives too large an error estimate.

3.4.1 Derivation of General Error Formulae

In the examples above, we saw that an error estimate ΔZ by the direct approach, i.e., by substituting $A + \Delta A$, $B + \Delta B$, $C + \Delta C$, ... directly into the formula $Z = Z(A, B, C, \ldots)$ leads to a cumbersome expression. We also stated that detailed statistical arguments show the error to be too large. A much better approach leading to 'neater' error formulae giving better error estimates in most cases is to use calculus, specifically the derivative (for $Z = Z(A)$) and the total differential (for $Z = Z(A, B, C, \ldots)$).

Consider first $Z = Z(A)$ where Z depends only on one measured quantity A. Then

$$\frac{dZ}{dA} = Z'(A).$$

Rearrangement gives

$$dZ = Z'(A)dA$$

Substituting the errors ΔZ and ΔA for the differentials immediately gives the required error formula:

$$\boxed{\text{ERROR IN } Z(A)} \quad \boxed{\Delta Z = \frac{dZ}{dA}\Delta A}$$

In some cases, relative errors are preferred. A general formula for relative errors for one-quantity formulae is immediately obtained by dividing the above formula by Z to get

$$\boxed{\text{RELATIVE ERROR IN } Z(A)} \quad \boxed{\frac{\Delta Z}{Z} = \frac{1}{Z(A)}\frac{dZ}{dA}\Delta A}$$

When Z depends on two measured quantities $Z = Z(A, B)$, we need to use the total differential

$$dZ = \frac{\partial Z}{\partial A}dA + \frac{\partial Z}{\partial B}dB$$

Again, substituting errors ΔZ, ΔA and ΔB for the differentials immediately gives the required error formula:

$$\boxed{\text{ERROR IN } Z(A, B)} \quad \boxed{\Delta Z = \frac{\partial Z}{\partial A}\Delta A + \frac{\partial Z}{\partial B}\Delta B}$$

Similarly, relative errors are preferred in most cases. Dividing the above formula by Z we get the required formula

$$\boxed{\text{RELATIVE ERROR IN } Z(A, B)} \quad \boxed{\frac{\Delta Z}{Z} = \frac{1}{Z}\frac{\partial Z}{\partial A}\Delta A + \frac{1}{Z}\frac{\partial Z}{\partial B}\wedge B}$$

Generalisation to more than two measured quantities is straightforward. The derivations that follow will be given for one and two-quantity formulae, while generalisations to more than two-quantity formulae will simply be stated.

3.4.2 One-Quantity Formula for Combining Errors

In the this section we use the one-quantity formula

$$\Delta Z = \frac{dZ(A)}{dA}\Delta A \tag{3.16}$$

or

$$\frac{\Delta Z}{Z} = \frac{1}{Z(A)}\frac{dZ(A)}{dA}\Delta A \tag{3.17}$$

to either derive an error and/or relative error formulae for various specific one-quantity formulae, whichever is appropriate. Note that the error ΔA may be a standard error in the mean or a systematic error.

Proportionalities and Inverse Proportionalities

For the proportionality

$$Z(A) = kA$$

formula (3.16) gives

$$\Delta Z = \frac{d(kA)}{dA} \Delta A = k \Delta A$$

The fractional error becomes

$$\frac{\Delta Z}{Z} = \frac{k \Delta A}{kA} = \frac{\Delta A}{A}$$

For the inverse proportionality

$$Z(A) = \frac{k}{A}$$

formula (3.16) gives

$$\Delta Z = \frac{d(kA^{-1})}{dA} \Delta A = -\frac{k}{A^2} \Delta A.$$

The fractional error becomes

$$\frac{\Delta Z}{Z} = \frac{k \Delta A}{A^2 k/A} = \frac{\Delta A}{A},$$

where the minus sign has been dropped since the error in an answer is given as $\pm \Delta Z$. We conclude that the required formula for both proportionalities and inverse proportionalities is:

ERROR FORMULA FOR PRORTIONALITIES AND INVERSE PROPORTIONALITIES	$\dfrac{\Delta Z}{Z} = \dfrac{\Delta A}{A}$	(3.18)

Powers

Now suppose Z is given by

$$Z(A) = kA^n, \qquad\qquad\qquad (3.19)$$

where $k = $ constant. Substituting

$$\frac{dZ}{dA} = knA^{n-1}$$

and Eq. (3.19) into Eq. (3.17) we get

$$\frac{\Delta Z}{Z} = \frac{knA^{n-1}}{kA^n} \Delta A = n \frac{\Delta A}{A}$$

The required formula is

$$\boxed{\text{ERROR FORMULA FOR POWERS} \quad \boxed{\frac{\Delta Z}{Z} = n\frac{\Delta A}{A}}} \qquad (3.20)$$

This formula applies even when n is negative. Though a negative error results, only the positive magnitude is taken since the final result is written $Z \pm \Delta Z$.

Exponentials

Consider the following formula containing an exponential:

$$Z(A) = ke^{nA} \qquad (3.21)$$

$$\frac{dZ}{dA} = nke^{nA} \qquad (3.22)$$

Substituting Eqs. (3.21) and (3.22) into Eq. (3.17) we get

$$\frac{\Delta Z}{Z} = \frac{nke^{nA}}{ke^{nA}}\Delta A = n\Delta A$$

Hence, the required formula is

$$\boxed{\text{ERROR FORMULA FOR EXPONENTIALS} \quad \boxed{\frac{\Delta Z}{Z} = n\Delta A}} \qquad (3.23)$$

Natural Logarithms

Consider the following formula containing natural logarithms:

$$Z = \ln A \qquad (3.24)$$

$$\frac{dZ}{dA} = \frac{1}{A} \qquad (3.25)$$

Substituting Eq. (3.25) into Eq. (3.16) we get the required formula:

$$\boxed{\text{ERROR FORMULA FOR NATURAL LOGARITHMS} \quad \boxed{\Delta Z = \frac{\Delta A}{A}}} \qquad (3.26)$$

Sines, Cosines and Tangents

Finally, consider the following trigonometric formulae and note that the angle θ is measured in radians:

$$Z_s(\theta) = \sin\theta,$$
$$Z_c(\theta) = \cos\theta,$$
$$Z_t(\theta) = \tan\theta.$$

Differentiating we get

$$\frac{dZ_s(\theta)}{d\theta} = \cos\theta$$

$$\frac{dZ_c(\theta)}{d\theta} = -\sin\theta$$

$$\frac{dZ_t(\theta)}{d\theta} = \sec^2\theta.$$

For trigonometric formulae the error and relative error formulae take similar forms so we will give both forms. Substituting the derivatives into Eqs. (3.16) and (3.17) we get

$$\boxed{\text{ERROR FORMULA FOR SINES}} \quad \boxed{\Delta Z = \cos\theta\,\Delta\theta, \quad \frac{\Delta Z}{Z} = \cot\theta\,\Delta\theta}$$

$$(3.27)$$

$$\boxed{\text{ERROR FORMULA FOR COSINES}} \quad \boxed{\Delta Z = -\sin\theta\,\Delta\theta, \quad \frac{\Delta Z}{Z} = -\tan\theta\,\Delta\theta}$$

$$(3.28)$$

$$\boxed{\text{ERROR FORMULA FOR TANGENTS}} \quad \boxed{\Delta Z = \sec^2\theta\,\Delta\theta, \quad \frac{\Delta Z}{Z} = \frac{\sec^2\theta}{\tan\theta}\,\Delta\theta}$$

$$(3.29)$$

3.4.3 Multi-quantity Formula for Combining Errors

For the derivation of specific multi-quantity error formulae we use the general formula

$$\Delta Z = \frac{\partial Z}{\partial A}\Delta A + \frac{\partial Z}{\partial B}\Delta B, \tag{3.30}$$

while for specific relative error formulae we use the general formula

$$\frac{\Delta Z}{Z} = \frac{1}{Z(A, B)}\frac{\partial Z}{\partial A}\Delta A + \frac{1}{Z(A, B)}\frac{\partial Z}{\partial B}\Delta B. \tag{3.31}$$

It is important to note that in the derivation of the error formulae the errors in the measured quantities should be interpreted as differences from the true values[1] A_0 and B_0, i.e., $\Delta A = A - A_0$ and $\Delta B = B - B_0$, and not as standard errors in the mean or as systematic errors. Once the formulae are derived, then they can be freely interpreted as standard errors in the mean, systematic errors or a combination of each.

Sums and Differences

Consider the sum of two measured quantities:

$$Z = kA + lB. \tag{3.32}$$

The partial derivatives are,

$$\frac{\partial Z}{\partial A} = k, \quad \frac{\partial Z}{\partial B} = l.$$

For sums (and differences) an error formula is preferred over a relative error formula. We can obtain an error formula by substituting the above derivatives into Eq. (3.30) to get

$$\Delta Z = k\Delta A + l\Delta B. \tag{3.33}$$

But, this formula gives too high an error estimate, since, as we mentioned above, some errors push the value down, whilst others push the value up, so that they have a tendency to cancel each other, reducing the overall error. A better error estimate is obtained by first squaring Eq. (3.33) to get

$$(\Delta Z)^2 = k^2(\Delta A)^2 + l^2(\Delta B)^2 + 2kl\Delta A\Delta B. \tag{3.34}$$

Next, interpret each term on the right hand side of Eq. (3.34) as the average error over many measured values. In this case, the first two terms are averages of positive quantities and are therefore nonzero. On the other hand, the third term is an average of an equal number of positive and negative terms and is therefore zero, i.e., $2kl\Delta A\Delta B = 0$. With these observations, Eq. (3.34) becomes

$$(\Delta Z)^2 = k^2(\Delta A)^2 + l^2(\Delta B)^2$$

Taking the square root of both sides gives the formula for a better error estimate:

---

[1]We recall that it is impossible to know the true value of a measured quantity. We can think of the true value as given by the mean of an infinite number of measurements. This assumption suggests that by taking enough readings we can approach the true value as closely as we please. Aside from the fact that taking an infinite number of measurements is impossible, the assumption that the mean of the infinite measurements gives the true value remains unproven. However, it is a good working hypothesis that the mean of many ideal measurements (measurements free of systematic error) is close to the true value.

$$\Delta Z = [k^2(\Delta A)^2 + l^2(\Delta B^2)]^{\frac{1}{2}}. \tag{3.35}$$

If instead, we are interested in the difference between two measured quantities

$$\Delta Z = k\Delta A - l\Delta B,$$

then

$$\frac{\partial Z}{\partial A} = k, \qquad \frac{\partial Z}{\partial B} = -l,$$

so that substituting into Eq. (3.30) and squaring gives

$$(\Delta Z)^2 = k^2(\Delta A)^2 + l^2(\Delta B)^2 - 2kl\Delta A\Delta B.$$

The third term is now negative, but since, for the same reason as for the sum, it is zero, taking the square root gives the same error formula (3.35) as for the sum. In general, when n quantities, $A_1, A_2, \ldots, A_n$, are added or subtracted to find a quantity Z, the error ΔZ in Z is given by

ERROR FOR SUMS DIFFERENCES, OR A MIXTURE OF THE TWO	$\Delta Z = [(\Delta A_1)^2 + (\Delta A_2)^2 + \ldots + (\Delta A_n)^2]^{\frac{1}{2}}$ (3.36)

Where constants multiply the measured quantities they can be inserted as in Eq. (3.35).

Products and Quotients

Suppose Z is given by

$$Z(A, B) = kAB.$$

Differentiating gives

$$\frac{\partial Z}{\partial A} = kB, \qquad \frac{\partial Z}{\partial B} = kA.$$

For products (and quotients), the relative error is preferred since it leads to a neater formula. Hence, we substitute the derivatives together with $Z(A, B) = kAB$ into Eq. (3.31) to get

$$\frac{\Delta Z}{Z} = \frac{1}{kAB}kB\Delta A + \frac{1}{kAB}kA\Delta B = \frac{\Delta A}{A} + \frac{\Delta B}{B}.$$

Once again, and for the same reason as for sums and differences, this formula gives an overly large error estimate. Following similar steps and argument as for sums and differences we get,

$$\left(\frac{\Delta Z}{Z}\right)^2 = \left(\frac{\Delta A}{A}\right)^2 + \left(\frac{\Delta B}{B}\right)^2 + 2\frac{\Delta A}{A}\frac{\Delta B}{B},$$

with $2\frac{\Delta A}{A}\frac{\Delta B}{B} = 0$. Taking the square root of both sides, we get a better relative error estimate formula:

$$\frac{\Delta Z}{Z} = \left[\left(\frac{\Delta A}{A}\right)^2 + \left(\frac{\Delta B}{B}\right)^2\right]^{\frac{1}{2}}. \tag{3.37}$$

Next, consider a multi-quantity quotient formula, and again aim for a formula for the relative error:

$$Z(A, B) = k\frac{A}{B}$$

$$\frac{\partial Z}{\partial A} = \frac{k}{B}, \qquad \frac{\partial Z}{\partial B} = -\frac{kA}{B^2}$$

$$\frac{\Delta Z}{Z} = \frac{1}{k\frac{A}{B}}\frac{k}{B}\Delta A - \frac{1}{k\frac{A}{B}}\frac{kA}{B^2}\Delta B = \frac{\Delta A}{A} - \frac{\Delta B}{B}$$

$$\left(\frac{\Delta Z}{Z}\right)^2 = \left(\frac{\Delta A}{A}\right)^2 + \left(\frac{\Delta B}{B}\right)^2 - 2\frac{\Delta A}{A}\frac{\Delta B}{B}$$

With the equation interpreted as an average over many measurements, the third term is again zero. Taking the square root of both sides, we get the desired better estimate formula:

$$\frac{\Delta Z}{Z} = \left[\left(\frac{\Delta A}{A}\right)^2 + \left(\frac{\Delta B}{B}\right)^2\right]^{\frac{1}{2}}.$$

It is identical to the relative error formula (3.35) for products. We may therefore write a general formula for n products or quotients of n measured quantities labelled by $A_1, A_2, \ldots, A_n$ as

ERROR FORMULA FOR PRODUCTS, QUOTIENTS OR A MIXTURE OF THE TWO	$\frac{\Delta Z}{Z} = \left[\left(\frac{\Delta A_1}{A_1}\right)^2 + \left(\frac{\Delta A_2}{A_2}\right)^2 + \cdots + \left(\frac{\Delta A_n}{A_n}\right)^2\right]^{\frac{1}{2}}$

$$\tag{3.38}$$

Powers

Consider the formula for a product of quantities raised to a power,

$$Z = kA^p B^q$$

We follow similar steps as above and note that a relative error formula is preferred:

$$\frac{\partial Z}{\partial A} = kpA^{p-1}B^q, \qquad \frac{\partial Z}{\partial B} = kqA^p B^{q-1}$$

$$\frac{\Delta Z}{Z} = \frac{1}{kA^p B^q}kpA^{p-1}B^q\Delta A + \frac{1}{kA^p B^q}kqA^p B^{q-1}\Delta B = p\frac{\Delta A}{A} + q\frac{\Delta B}{B}$$

The resulting relative error formula is similar to that for products and quotients. By similar reasoning and mathematical steps, a better relative error formula is found to be

$$\frac{\Delta Z}{Z} = \left[p^2 \left(\frac{\Delta A}{A} \right)^2 + q^2 \left(\frac{\Delta B}{B} \right)^2 \right]^{\frac{1}{2}}.$$

The formula is easily generalised for the case $Z = kA_1^{m_1} A_2^{m_2} \dots A_n^{m_n}$:

ERROR FORMULA FOR POWERS	$\dfrac{\Delta Z}{Z} = \left[m_1^2 \left(\dfrac{\Delta A_1}{A_1} \right)^2 + m_2^2 \left(\dfrac{\Delta A_2}{A_2} \right)^2 + \dots + m_n^2 \left(\dfrac{\Delta A_n}{A_n} \right)^2 \right]^{\frac{1}{2}}$

$$(3.39)$$

3.4.4 An Example of the Use of the Combination Formulae

Consider the formula

$$\eta = \frac{\rho \pi r^4}{8l Q}.$$

We want to find the error $\Delta \eta$ in η due to the errors $\Delta \rho$, Δr, Δl and ΔQ in the measured quantities ρ, r, l and Q.

We can use the formula for products and quotients by first setting $r^4 = B$,

$$\eta = \frac{\rho \pi B}{8l Q}.$$

The relative error is

$$\frac{\Delta\eta}{\eta} = \left[\left(\frac{\Delta\rho}{\rho}\right)^2 + \left(\frac{\Delta l}{l}\right)^2 + \left(\frac{\Delta Q}{Q}\right)^2 + \left(\frac{\Delta B}{B}\right)^2\right]^{\frac{1}{2}}.$$

Notice that the constants 8 and π make no contribution to the error. Then find $\Delta B/B$ in terms of r using Eq. (3.21):

$$B = r^4$$
$$\frac{\Delta B}{B} = 4\frac{\Delta r}{r}.$$

Hence,

$$\frac{\Delta\eta}{\eta} = \left[\left(\frac{\Delta\rho}{\rho}\right)^2 + \left(\frac{\Delta l}{l}\right)^2 + \left(\frac{\Delta Q}{Q}\right)^2 + \left(\frac{4\Delta r}{r}\right)^2\right]^{\frac{1}{2}}.$$

3.5   Common Sense with Errors

Sometimes you may find that when you measure a number of quantities the error in one of them is much smaller than the error in the other quantities. When squares are taken the difference is increased still further.

Consider the following example:

$$Z = A + B$$

with

$$A = (100 \pm 10) \quad \text{and} \quad B = (100 \pm 3).$$

Then

$$\Delta A = 10, \quad \text{and} \quad \Delta B = 3,$$

so that

$$\Delta Z = [(\Delta A)^2 + (\Delta B)^2]^{\frac{1}{2}}$$
$$\Delta Z = [(10)^2 + (3)^2]^{\frac{1}{2}}$$
$$\Delta Z = (100 + 9)^{\frac{1}{2}} = 109^{\frac{1}{2}}$$
$$\Delta Z = 10.4$$

Neglecting the error in B we get

$$\Delta Z = [(\Delta A)^2]^{\frac{1}{2}}$$
$$\Delta Z = \Delta A$$
$$\Delta Z = 10$$

We see that the error of $\Delta B = 3$ contributes only 0.4 to the error $\Delta Z = 10.4$ in Z, i.e., makes only a 3.8% contribution. It would be a reasonable approximation therefore to neglect the contribution ΔB in calculating the error in Z.

We conclude that when a number of quantities are measured, we may neglect the contribution to the total error of those measured quantities which have a small error (with care!).

3.6 Proportionalities

A great deal of experimental work is concerned with how one quantity changes as a result of another. There are several types of such dependence:

1. When
$$x = kt,$$

   with k constant, we say that x is proportional to t and write

$$x \propto t.$$

   Or when
$$s = kt^2,$$

   with k constant, we say that s is proportional to t^2 and write

$$s \propto t^2.$$

2. When
$$\rho = k\frac{1}{v},$$

   with k constant, we say that ρ is inversely proportional to v.
3. When
$$\rho = \frac{k\mu}{v},$$

   k constant, we say that ρ is proportional to μ and inversely proportional to v.

3.7   Graphs

Graphs are used because they make it easy to see how one quantity is related to another. For example, a graph can show whether or not two quantities are proportional, or if a point does not fit a general pattern (which indicates a bad measurement).

3.7.1   *Straight-Line Graphs. Determination of Slope and Intercept*

In many situations it is often possible to plot a straight-line graph. To see whether a graph should be a straight line, compare the formula which relates the quantities you are plotting (which you will often, but not always, know beforehand) to the equation of a straight line. The equation of a straight line is

$$y = mx + c,$$

where m and c are constants. $m =$ slope of the straight line; $c =$ where the line cuts the y-axis (Fig. 3.5).

NOTE 1. If $c = 0$ then

$$y = mx.$$

The line cuts the x-axis when $y = 0$, i.e., at $x = 0$. The line therefore passes through the origin.

NOTE 2. The line cuts the x-axis when $y = 0$, i.e., at

$$0 = mx + c$$
$$x = \frac{-c}{m}.$$

Example 3.7.1 (To test whether or not the graph to be plotted is a straight line graph)
Suppose we want to plot ρ against μ (ρ v. μ), where

$$\rho = k\mu, \qquad k = \text{const.}$$

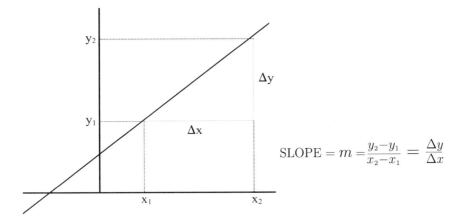

Fig. 3.5 The slope of a straight-line graph

$$\text{SLOPE} = m = \frac{y_2 - y_1}{x_2 - x_1} = \frac{\Delta y}{\Delta x}$$

Comparing with

$$y = mx + c$$

we see that $m = k$ and $c = 0$ and we should expect a straight line of slope m passing through the origin.

Now, suppose you have plotted 12 points on an xy-graph and the points are equally spaced along the x-axis as shown in Fig. 3.6. What is the best line through the points? One method you can use to draw the best straight line, which also provides a method for finding the error in the slope, is the POINTS-IN-PAIRS method. Another common and important method, but which involves much more work is the METHOD OF LEAST SQUARES. We describe both below.

3.7.2 The Points-In-Pairs Method

THE POINTS-IN-PAIRS METHOD. POINTS ALONG x-AXIS EQUALLY SPACED

Refer to Fig. 3.6 for what follows.

FIRST: PLOT AN EVEN NUMBER OF POINTS (12 IN OUR EXAMPLE).

SECOND: DIVIDE THE POINTS INTO TWO EQUAL GROUPS BY A
                    VERTICAL DOTTED LINE.

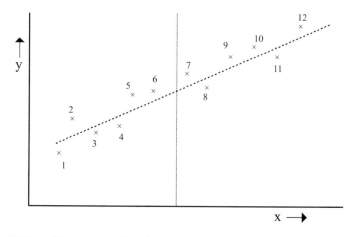

Fig. 3.6 Division of data points in the points-in-pairs method

THIRD: PAIR THE POINTS OFF AS FOLLOWS:

1 and 7, 2 and 8, 3 and 9, . . . , 6 and 12

FOURTH: CALCULATE THE DIFFERENCE IN y-VALUES FOR EACH
                       PAIR:

$$(y_7 - y_1), \ (y_8 - y_2), \ (y_9 - y_3), \ldots, \ (y_{12} - y_6)$$

FIFTH: CALCULATE THE MEAN OF THESE DIFFERENCES:

$$\overline{D_y} = \frac{(y_7 - y_1) + (y_8 - y_2) + \ldots + (y_{12} - y_6)}{6}$$

SIXTH: SINCE THE POINTS ON THE x-AXIS ARE EQUIDISTANT, THE
                       DIFFERENCE IN x-VALUES OF THE PAIRS IS THE SAME:

$$D_x = (x_7 - x_1) = (x_8 - x_2) = \ldots = (x_{12} - x_6)$$

SEVENTH: DETERMINE THE MEAN SLOPE USING

$$\boxed{\text{MEAN SLOPE}} \quad \boxed{\overline{m} = \frac{\overline{D_y}}{D_x}}$$

EIGHTH: DETERMINE THE MEAN VALUES OF $\overline{x}$ AND $\overline{y}$:

$$\overline{x} = \frac{x_1 + x_2 + \ldots + x_{12}}{12}, \quad \overline{y} = \frac{y_1 + y_2 + \ldots + y_{12}}{12}$$

Then,

> THE BEST STRAIGHT LINE BY THIS METHOD IS THE ONE
> WITH SLOPE $\overline{m} = \frac{D_y}{D_x}$, PASSING THROUGH THE POINT $(\overline{x}, \overline{y})$.

THE 'POINTS-IN-PAIRS' METHOD. POINTS ALONG x-AXIS NOT EQUALLY SPACED

Note: As far as possible, the variation in the distance between the points on the x-axis should not be too great (since the slopes obtained from the more widely spaced points would give better estimates than those based on closer points).

FIRST TO FOURTH STEPS ARE THE SAME AS ABOVE.

FIFTH: CALCULATE THE DIFFERENCE IN x-VALUES FOR EACH
            PAIR:

$$(x_7 - x_1), \ (x_8 - x_2), \ (x_9 - x_3), \ldots, \ (x_{12} - x_6)$$

SIXTH: DETERMINE THE SLOPE FOR EACH x, y PAIR:

$$m_1 = \frac{y_7 - y_1}{x_7 - x_1}, \ m_2 = \frac{y_8 - y_2}{x_8 - x_2}, \ldots, \ m_6 = \frac{y_{12} - y_6}{x_{12} - x_6}$$

SEVENTH: DETERMINE THE MEAN SLOPE:

$$\overline{m} = \frac{m_1 + m_2 + \ldots + m_6}{6}$$

EIGHTH: DETERMINE THE MEAN VALUES OF $\overline{x}$ AND $\overline{y}$:

$$\overline{x} = \frac{x_1 + x_2 + \ldots + x_{12}}{12}, \quad \overline{y} = \frac{y_1 + y_2 + \ldots + y_6}{12}$$

Then,

> THE BEST STRAIGHT LINE BY THIS METHOD IS THE ONE
> WITH SLOPE $\overline{m}$ PASSING THROUGH THE POINT $(\overline{x}, \overline{y})$.

Standard Error in the Slope

EQUAL SPACING ON x-AXIS

Calculate the slope for each pair of points:

$$m_1 = \frac{y_7 - y_1}{D_x}, \quad m_2 = \frac{y_8 - y_2}{D_x}, \ldots, \quad m_6 = \frac{y_{12} - y_6}{D_x}.$$

Use the mean of the slope m already calculated for the case of equal x-axis spacing, i.e.,

$$\overline{m} = \frac{\overline{D_y}}{D_x}$$

UNEQUAL SPACING ON x-AXIS

Use the mean slope $\overline{m}$ already calculated for the case of unequal x-axis spacing, i.e.,

$$\overline{m} = \frac{m_1 + m_2 + \ldots + m_6}{6}$$

FROM NOW ON THE PROCEDURE IS THE SAME FOR EITHER CASE

CALCULATE RESIDUALS:

$$d_1 = m_1 - \overline{m}$$
$$d_2 = m_2 - \overline{m}$$
$$\vdots \quad \vdots \quad \vdots$$
$$d_3 = m_6 - \overline{m}$$

CALCULATE THE STANDARD DEVIATION:

$$\sigma = \left(\frac{d_1^2 + d_2^2 + d_3^2 + \cdots d_6^2}{6} \right)^{\frac{1}{2}}$$

Then,

STANDARD ERROR IN THE SLOPE	$s_m = \dfrac{\sigma}{(n-1)^{\frac{1}{2}}}$

We need to add some additional comments for lines passing through the origin and those which do not.

GRAPH $y = mx$, $(c = 0)$

From the equation of a straight line we know that the line must pass through the origin. Nevertheless, the 'points-in-pairs' method must be used as usual, even though the 'best-line' may not pass through the origin. One reason is that the amount by which the line misses the origin may contain information about a possible systematic error in the apparatus.

GRAPH $y = mx + c$

Finding the point c where the 'best-line' cuts the y-axis is straightforward when you can plot the graph to include the origin. You can simply read-off c from the graph (Fig. 3.7).

Sometimes, however, plotting a graph which includes the origin requires a choice of scale which results in the plotted points being too close together (see Fig. 3.8). This makes plotting an accurate line difficult and it is much better to choose a different scale which spreads the points along the x-axis (see Fig. 3.9). The problem now is that c cannot be simply read-off the graph. In this case, c can be easily calculated by using the slope $\overline{m}$ of the 'best-line' and the means $\overline{x}$ and $\overline{y}$:

$$c = \overline{y} - \overline{m}\,\overline{x}.$$

Fig. 3.7 With axis scales
that include the origin, c can
be read off directly

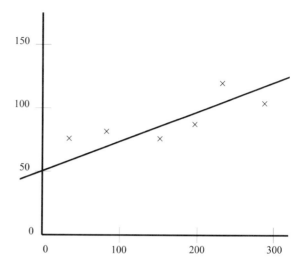

Fig. 3.8 Points are too close
together. A change of scale is
needed

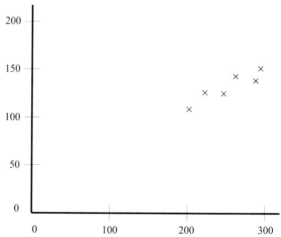

3.7.3   Logarithmic Graph Paper

Suppose you have an expression

$$y = kA^x, \tag{3.40}$$

where k and A are constants. Plotting y against x will not give a straight line. Since
dealing with straight lines is easier, it is desirable to express Eq. (3.40) in such a way
that a straight-line graph can be given. This can be done using logarithms.

Recall the following rules for logarithms:

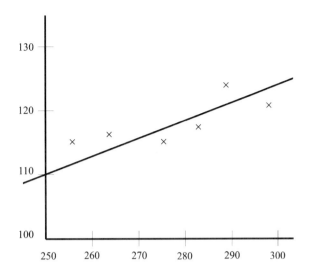

Fig. 3.9 To spread the points adequately, a scale is chosen which excludes the origin

1. $\log (ab) = \log a + \log b$
2. $\log (a/b) = \log a - \log b$
3. $\log a^n = n \log a$

Take logs of both sides of Eq. (3.40)

$$\log y = \log k A^x$$
$$\log y = \log A^x + \log k$$
$$\log y = x \log A + \log k$$
$$\log y = (\log A)x + \log k$$

Comparing this with the equation of a straight line, $y = mx + c$, we see that plotting $\log y$ against x results in a straight line with slope $m = \log A$ which cuts the y-axis at $c = \log k$.

Though we can always look up the logs of our y-readings and plot the graph on ordinary graph paper, it is much more convenient to use special logarithmic graph paper, where the scale along the y-axis is logarithmic.

Sometimes, to get a straight line you must plot $\log y$ against $\log x$. Again there is logarithmic paper where both the x and y-axes are logarithmic.

3.7.4 Transformation to Linear Graphs

In a similar fashion to the previous subsection, other relations between quantities can be transformed to a linear form and hence plotted as a linear graph. Below is a summary of linear forms corresponding to various formula and the quantities to be plotted to produce a linear graph.

Relation	Linear form	Plot:
$y = mx + c$	$y = mx + c$	y v. x
$y = Ax^n$	$\log y = n \log x + \log A$	$\log y$ v. $\log x$
$y = A(x + x_0)^n$	$y^{\frac{1}{n}} = A^{\frac{1}{n}}x + A^{\frac{1}{n}}x_0$	$y^{\frac{1}{n}}$ v. x
$y = Ax^3 + cx$	$\frac{y}{x} = Ax^2 + c$	$\frac{y}{x}$ v. x^2
$y = y_0 e^{\frac{-x}{x_0}}$	$\log_e y = \log_e y_0 - \frac{x}{x_0}$	$\log_e y$ v. x

3.8   Percentage Error

The errors we have been calculating can be easily converted to percentage errors. If ΔZ is the error in a quantity Z, then the percentage error is given by

$$\text{PERCENTAGE ERROR, } e_p \qquad\qquad e_p = \frac{\Delta Z}{Z} \times 100$$

In other words, the percentage error is just the relative error multiplied by 100.

3.9   Problems

1. The diameter of a wire is measured repeatedly in different places along its length. The measurements in millimetres are

    1.26,   1.26,   1.29,   1.31 , 1.28,   1.27,   1.26,   1.25,
    1.28,   1.32,   1.21,   1.22,   1.29,   1.28,   1.27.

    (a) Calculate the standard deviation of these readings.
    (b) Calculate the standard error.

2. Given the following table of values, calculate the slope $\overline{m}$ and $(\overline{x}, \overline{y})$ for the best straight line using the points-in-pairs method. Also calculate where the line cuts the y-axis, and the standard error in the slope.

$$x = 2.0 \ \ 4.0 \ \ 6.0 \ \ 8.0 \ \ 10.0 \ \ 12.0$$
$$y = 7.1 \ \ 8.2 \ \ 8.8 \ \ 10.0 \ \ 11.0 \ \ 11.8$$

3. Two objects have weights (100 ± 0.4) g and (50 ± 0.3) g.

   (a) Calculate the sum of their weights and the error in the sum of their weights. Write the sum together with its error to the correct number of significant figures.
   (b) Calculate the difference of their weights and the error in the difference of their weights. Write the difference together with its error to the correct number of significant figures.

4. Three objects have weights (100 ± 0.4) g, (50 ± 0.3) g and (200 ± 0.5) g. Calculate the sum of the weights and the error in the sum. Write your answer together with its error to the correct number of significant figures.

5. The volume of a rectangular block is calculated from the following measurements of its dimensions: (10.00 ± 0.10) cm, (5.00 ± 0.05) cm and (4.00 ± 0.04) cm. Calculate the volume and the error in the volume. Write your answer to the correct number of significant figures.

6. The pressure of a gas can be estimated from the force exerted on a given area. If the force is (20.0 ± 0.5) N, and the area is rectangular with sides (5.0 ± 0.2) mm and (10.0 ± 0.5) mm, what is the relative error in the value of the pressure? What is the percentage error in the value of the pressure?

7. Calculate the area and the error in the area of a circle whose radius is determined to be (14.6 ± 0.5) cm. Write your answer to the correct number of significant figures.

Chapter 4
The Method of Least Squares

The method of least squares is a standard statistical method for drawing the best straight line through a set of points.

Consider n pairs of measurements $(x_1, y_1), (x_2, y_2), \ldots, (x_n, y_n)$. We assume the errors are entirely in the y value. The analysis when there are errors in both the x and y values is much more complicated, yet with little gain in accuracy. Hence, we shall confine ourselves to the former case, which is what obtains in practically all cases. We will also assume that each pair has equal weight. We see from the Fig. 4.1 that the d_i given by

$$d_i = y_i - Y_i$$

are the deviations of the measured values y_i from the value Y_i given by the best line (as yet unknown) through the data.

Now, the Y_i are given by

$$Y_i = mx_i + c$$

where m is the slope of the unknown best line and c is where the unknown best line cuts the y-axis.

The best values of m and c are taken to be those for which the sum of the deviations squared,

$$S = \sum_{i=1}^{i=n} d_i^2 = \sum_{i=1}^{i=n} (y_i - mx_i - c)^2, \tag{4.1}$$

© Springer International Publishing AG, part of Springer Nature 2018
P. N. Kaloyerou, *Basic Concepts of Data and Error Analysis*,
https://doi.org/10.1007/978-3-319-95876-7_4

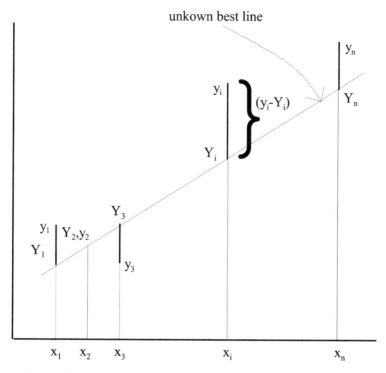

Fig. 4.1 Method of least squares

is a minimum - hence the name "method of least squares". This method was suggested by Gauss[1] in 1801 and Legendre[2] in 1806.

---

[1] Carl Friedrich Gauss (1777–1855), originally named JohanFriedrich Carl Gauss, was a German mathematician regarded as one of the great mathematicians of his time, making contributions to number theory, geometry, probability theory, cartography, terrestrial magnetism, orbital astronomy, theory of functions and potential theory (a branch of mathematics arising from electromagnetism and gravitation). In relation to his work on cartography, he invented the heliotrope (an instrument that focuses sunlight into a beam that can be seen from several miles away) and in his work on terrestrial magnetism he invented the magnetometer to measure the Earth's magnetic field. With his Göttingen colleague, Wilhelm Weber, he made the first electric telegraph. Over many years, he gave four derivations of the fundamental theorem of algebra. His two most important contributions were published in 1801. The first publication was *Disquisitiones Arithmeticae* (*Algebraic Number Theory*) and the second concerned the rediscovery of the asteroid Ceres. The importance of the work on the Ceres asteroid has to do with the development of an ingenious method for dealing with errors in observations, the method of least squares. Some years later, Legendre produced a comprehensive exposition of this method (see footnote 2), and both Gauss and Legendre came to share credit for the method.

[2] Adrien-Marie Legendre (1752–1833) was born in Paris, France. He was a French mathematician who made contributions in many areas of mathematics including number theory, geometry, mechanics, orbital astronomy, and elliptic integrals. Legendre was appointed, with Cassini and Mechain, to a special committee to develop the metric system and, in particular, to make measurements to

As usual, the minimum is found by setting the derivative equal to zero. Differentiating Eq. (4.1) with respect to m and equating the result to zero gives

$$\frac{\partial S}{\partial m} = \sum_{i=1}^{n} -2x_i (y_i - mx_i - c) = 0$$

$$= -2 \left[\sum_{i=1}^{n} x_i y_i - m \sum_{i=1}^{n} x_i^2 - c \sum_{i=1}^{n} x_i \right] = 0.$$

This implies that

$$m \sum_{i=1}^{n} x_i^2 + c \sum_{i=1}^{n} x_i = \sum_{i=1}^{n} x_i y_i. \tag{4.2}$$

Also, differentiating Eq. (4.1) with respect to c and equating the result to zero gives

$$\frac{\partial S}{\partial c} = -2 \sum_{i=1}^{n} (y_i - mx_i - c) = 0$$

$$= -2 \left[\sum_{i=1}^{n} y_i - m \sum_{i=1}^{n} x_i - nc \right] = 0.$$

This implies that

$$\sum_{i=1}^{n} y_i = m \sum_{i=1}^{n} x_i + nc.$$

Dividing by n we get

$$\sum_{i=1}^{n} \frac{y_i}{n} = m \sum_{i=1}^{n} \frac{x_i}{n} + c. \tag{4.3}$$

With

$$\overline{y} = \sum_{i=1}^{n} \frac{y_i}{n}, \qquad \overline{x} = \sum_{i=1}^{n} \frac{x_i}{n}$$

we get

$$\overline{y} = m\overline{x} + c. \tag{4.4}$$

determine the **standard metre**. He also worked on projects to produce logarithmic and trigonometric tables. His work on elliptic integrals, regarded as his most important contribution, was published in his treatise *Traité des fonctions elliptiques* (*Treatise on Elliptic Functions*), 1825–37. The first comprehensive treatment of the *method of least squares* is contained in Legendre's 1806 book *Nouvelles méthodes pour la détermination des orbites des comètes* (*New Methods for the Determination of Comet Orbits*) and shares credit for its discovery with his German rival Carl Friedrich Gauss.

The best straight line passes through the point $(\bar{x}, \bar{y})$, called the CENTROID. Solving Eqs. (4.2) and (4.3) simultaneously, we obtain (see Sect. 4.1 for the proof).

SLOPE OF THE BEST STRAIGHT LINE	$m = \dfrac{\sum_{i=1}^{n}(x_i - \bar{x})y_i}{\sum_{i=1}^{n}(x_i - \bar{x})^2}$	(4.5)

WHERE THE BEST LINE CUTS THE y-AXIS	$c = \bar{y} - m\bar{x}$	(4.6)

CAUTION: Equation (4.4) cannot be used to determine the slope m. To determine a straight line, either two points through which it passes must be given, or one point and a slope m must be given. Equation (4.4) contains only a single point, the centroid $(\bar{x}, \bar{y})$, an unknown slope m and an unknown c, and hence cannot be used to determine the slope m. Put another way, there is an infinite number of straight lines passing through the centroid $(\bar{x}, \bar{y})$ with different slopes (with, correspondingly, an infinite number of values of c) consistent with Eq. (4.4). Hence, it cannot be used to determine the slope of the best straight line. Once the slope of the best straight line is determined using Eq. (4.5), then using the calculated centroid $(\bar{x}, \bar{y})$, the constant c can be determined using Eq. (4.4) or (4.6).

4.1 Proof of Equation (4.5)

From Eq. (4.2) we get

$$c \sum_{i=1}^{n} x_i = \sum_{i=1}^{n} x_i y_i - m \sum_{i=1}^{n} x_i^2.$$

Divide by n:

$$c \sum_{i=1}^{n} \frac{x_i}{n} = \sum_{i=1}^{n} \frac{x_i y_i}{n} - m \sum_{i=1}^{n} \frac{x_i^2}{n}.$$

Substitute $\sum_{i=1}^{n} \dfrac{x_i}{n} = \bar{x}$ and rearrange:

$$c = \frac{1}{\bar{x}} \left(\sum_{i=1}^{n} \frac{x_i y_i}{n} - m \sum_{i=1}^{n} \frac{x_i^2}{n} \right). \tag{4.7}$$

From Eq. (4.3) we get

$$c = \sum_{i=1}^{n} \frac{y_i}{n} - m \sum_{i=1}^{n} \frac{x_i}{n}. \tag{4.8}$$

Equate Eqs. (4.7) and (4.8) and rearrange:

$$\sum_{i=1}^{n} \frac{y_i}{n} - m \sum_{i=1}^{n} \frac{x_i}{n} = \frac{1}{\bar{x}} \left(\sum_{i=1}^{n} \frac{x_i y_i}{n} - m \sum_{i=1}^{n} \frac{x_i^2}{n} \right),$$

$$m \sum_{i=1}^{n} \frac{x_i^2}{n} - m \sum_{i=1}^{n} \frac{x_i}{n} \bar{x} = \sum_{i=1}^{n} \frac{x_i y_i}{n} - \sum_{i=1}^{n} \frac{y_i}{n} \bar{x},$$

$$\frac{m}{n} \sum_{i=1}^{n} \left(x_i^2 - x_i \bar{x} \right) = \frac{1}{n} \sum_{i=1}^{n} (x_i - \bar{x}) y_i.$$

From this we get the formula for the slope

$$m = \frac{\sum_{i=1}^{n} (x_i - \bar{x}) y_i}{\sum_{i=1}^{n} \left(x_i^2 - x_i \bar{x} \right)}. \qquad (4.9)$$

The denominator can be transformed into a neater form as follows:

$$\sum_{i=1}^{n} \left(x_i^2 - x_i \bar{x} \right) = \sum_{i=1}^{n} x_i^2 - \bar{x} \left(\sum_{i=1}^{n} \frac{x_i}{n} \right) n = \sum_{i=1}^{n} x_i^2 - \bar{x}^2 n.$$

Add and subtract $n\bar{x}^2$ on the right-hand side to get

$$\sum_{i=1}^{n} \left(x_i^2 - x_i \bar{x} \right) = \sum_{i=1}^{n} x_i^2 - 2\bar{x}^2 n + n\bar{x}^2,$$

$$= \sum_{i=1}^{n} x_i^2 - 2\bar{x} \left(\sum_{i=1}^{n} \frac{x_i}{n} \right) n + n\bar{x}^2,$$

$$= \sum_{i=1}^{n} \left(x_i^2 - 2\bar{x} x_i + \bar{x}^2 \right),$$

$$= \sum_{i=1}^{n} (x_i - \bar{x})^2.$$

Substituting this result into Eq. (4.9) we get Eq. (4.5):

$$m = \frac{\sum_{i=1}^{n} (x_i - \bar{x}) y_i}{\sum_{i=1}^{n} (x_i - \bar{x})^2},$$

which completes the proof.

4.2   Standard Error in m and c of Best Straight Line

Let Δm be the standard error in the slope, and Δc be the standard error in c. We state without proof that estimates of the standard errors are given by

STANDARD ERROR IN THE SLOPE	$\Delta m = \left[\dfrac{\sum_{i=1}^{n} d_i^2}{D(n-2)} \right]^{\frac{1}{2}}$	(4.10)

STANDARD ERROR IN c	$\Delta c = \left[\left(\dfrac{1}{n} + \dfrac{\bar{x}^2}{D} \right) \cdot \dfrac{\sum_{i=1}^{n} d_i^2}{(n-2)} \right]^{\frac{1}{2}}$	(4.11)

where

The sum of the squares of the residuals of the $x-$values	$D = \sum_{i=1}^{n} (x_i - \bar{x})^2$	(4.12)

and

The difference between y_i and $Y_i = mx_i + c$, where m and c are the best values	$d_i = y_i - Y_i = y_i - mx_i - c$	(4.13)

Proofs of Eqs. (4.10) and (4.11) can be found in Appendix C of G.L. Squires, *Practical Physics*, fourth edition, (Cambridge University Press, Cambridge, 2001).

4.3   Lines Through the Origin

If we require the best straight line to pass through the origin, i.e., $c = 0$, the best value of m is given by setting $c = 0$ in Eq. (4.2),

$$m = \frac{\sum_{i=1}^{n} x_i y_i}{\sum_{i=1}^{n} x_i^2}$$

(4.14)

An estimate of the standard error is given by

$$\Delta m = \left[\frac{1}{(n-1)} \cdot \frac{\sum_{i=1}^{n} d_i^2}{\sum_{i=1}^{n} x_i^2} \right]^{\frac{1}{2}}$$

(4.15)

where Y_i reduces to $Y_i = mx_i$ so that d_i reduces to

$$(4.16)$$

$$d_i = y_i - Y_i = y_i - mx_i$$

Even though we know that the line passes through the origin, it can be useful to use Eq. (4.5) to determine the slope m of the best straight line, and Eq. (4.6) to determine the constant c where it cuts the $y-$axis. In this case the distance by which the best line misses the origin may give a visual indication of the error.

4.4 Summary

1.

$$\text{mean} = \bar{x} = \frac{x_1 + x_2 + \ldots + x_n}{n}$$ $n = $ number of data points

2.

RESIDUALS

$$d_1 = x_1 - \bar{x}$$
$$d_2 = x_2 - \bar{x}$$
$$\vdots \quad \vdots \quad \vdots$$
$$d_n = x_n - \bar{x}$$

3.

STANDARD DEVIATION

$$\sigma = \left(\frac{d_1^2 + d_2^2 + d_3^2 + \cdots d_n^2}{n} \right)^{\frac{1}{2}}$$

4.

STANDARD ERROR IN THE MEAN

$$s_m = \frac{\sigma}{(n-1)^{\frac{1}{2}}}$$

5. ERROR IN PROPORTIONALITIES AND INVERSE PROPORTIONALITIES

ERROR FORMULA FOR PRORTIONALITIES AND INVERSE PROPORTIONALITIES	$\dfrac{\Delta Z}{Z} = \dfrac{\Delta A}{A}$

6. ERROR IN POWERS

$$Z = kA^n$$

$$\boxed{\dfrac{\Delta Z}{Z} = \dfrac{n\Delta A}{A}}$$

7. ERROR IN NATURAL LOGARITHMS

$$Z = \ln A$$

$$\boxed{\Delta Z = \dfrac{\Delta A}{A}}$$

8. ERROR IN EXPONENTIALS

$$Z(A) = ke^{nA}$$

$$\boxed{\dfrac{\Delta Z}{Z} = n\Delta A}$$

9. ERROR IN SINES, COSINES AND TANGENTS

$$
\begin{aligned}
Z = \sin\theta &= \Delta Z = \cos\theta\,\Delta\theta, & \frac{\Delta Z}{Z} &= \cot\theta\,\Delta\theta \\
Z = \cos\theta &= \Delta Z = -\sin\theta\,\Delta\theta, & \frac{\Delta Z}{Z} &= -\tan\theta\,\Delta\theta \\
Z = \tan\theta &= \Delta Z = \sec^2\theta\,\Delta\theta, & \frac{\Delta Z}{Z} &= \frac{\sec^2\theta}{\tan\theta}\,\Delta\theta
\end{aligned}
$$

10. ERROR IN SUMS AND DIFFERENCES

$$Z = A_1 + A_2 + \cdots + A_n$$

$$\Delta Z = [(\Delta A_1)^2 + (\Delta A_2)^2 + \cdots + (\Delta A_n)^2]^{\frac{1}{2}}$$

11. ERROR IN PRODUCTS AND QUOTIENTS

$$Z = \frac{A_1 A_2 A_3 \ldots A_n}{B_1 B_2 B_3 \ldots B_m}$$

$$\frac{\Delta Z}{Z} = \left[\left(\frac{\Delta A_1}{A_1} \right)^2 + \left(\frac{\Delta A_2}{A_2} \right)^2 + \cdots + \left(\frac{\Delta A_n}{A_n} \right)^2 \right.$$
$$\left. + \left(\frac{\Delta B_1}{B_1} \right)^2 + \left(\frac{\Delta B_2}{B_2} \right)^2 + \cdots + \left(\frac{\Delta B_m}{B_m} \right)^2 \right]^{\frac{1}{2}}$$

12. METHOD OF LEAST SQUARES

SLOPE OF THE BEST STRAIGHT LINE	$m = \dfrac{\sum_{i=1}^{n}(x_i - \overline{x}) y_i}{\sum_{i=1}^{n}(x_i - \overline{x})^2}$

WHERE BEST LINE CUTS y−AXIS	$c = \overline{y} - m\overline{x}$

STANDARD ERROR IN THE SLOPE	$\Delta m = \left[\dfrac{\sum_{i=1}^{n} d_i^2}{D(n-2)} \right]^{\frac{1}{2}}$

STANDARD ERROR IN c	$\Delta c = \left[\left(\dfrac{1}{n} + \dfrac{\overline{x}^2}{D} \right) \cdot \dfrac{\sum_{i=1}^{n} d_i^2}{(n-2)} \right]^{\frac{1}{2}}$

where

The sum of the squares of the residuals of the x–values	$D = \sum_{i=1}^{n}(x_i - \bar{x})^2$

The difference between y_i and $Y_i = mx_i + c$, where m and c are the best values	$d_i = y_i - Y_i = y_i - mx_i - c$

13. PERCENTAGE ERROR

PERCENTAGE ERROR, e_p	$e_p = \dfrac{\Delta Z}{Z} \times 100$

Chapter 5
Theoretical Background - Probability and Statistics

So far we have presented the essential elements of statistics needed for data and error analysis. For a deeper understanding of the concepts and formulae presented in the first four chapters it is necessary to consider essential elements of probability theory as well as a more complete treatment of statistics. This then is the purpose of this chapter. To achieve a coherent self-contained treatment there may be a little repetition of earlier material. This chapter can be viewed as a stand-alone chapter and can be read independently of the other chapters. To begin writing up experiments, only chapters one to four are needed. Chapter 6 introduces computer methods and can also be read independently after chapters one to four have been mastered. Indeed, it is highly recommended that students learn to do calculations with a calculator and draw graphs by hand before moving on to computers.

Below we give a brief history of the emergence of probability, but we will first start with interpretations of probability. The reason for this ordering is that there are two classes of interpretation of probability which permeate the development of probability and the originators of probability swayed from one interpretation to another, even where they stated a preference for only one of the interpretations. The two classes of interpretations in question are the *objective interpretation* and the *subjective interpretation*. Knowing these two interpretations will help in understanding the essential elements of probability and statistics, which it is the purpose of this chapter to present.

5.1 Introduction

Probability and statistics arise in almost every human endeavor: the sciences, medicine, sociology, politics, insurance, games, gambling and so on. Many conclusions in medicine on what is good or bad for us result from statistical studies where a large sample of people is studied under specific conditions (a particular diet

© Springer International Publishing AG, part of Springer Nature 2018 71
P. N. Kaloyerou, *Basic Concepts of Data and Error Analysis*,
https://doi.org/10.1007/978-3-319-95876-7_5

or a particular medicine). Though, here, we are specifically interested in scientific experiments, the concepts of this chapter can be applied quite generally.

We should mention that there is an essential difference between what is called *classical probability* and what is called *quantum probability*. Classical probability concerns all probabilities that arise outside of the quantum theory. We can characterise classical probability as arising because of a lack of complete knowledge, but that in principle, if not in practice, the probability aspect can be eliminated with complete knowledge. As an example, think of tossing a (fair) coin. Before the coin lands we don't know whether it will be heads-up or tails-up, but instead attribute a probability of 0.5 of getting heads and similarly for tails. But if, though extremely difficult, if not impossible, we measure the exact initial orientation and position of the coin as well as the exact magnitude and direction of the applied force in tossing the coin, we can use the laws of mechanics to exactly predict the outcome of tossing the coin.

In classical physics theories (mechanics, hydrodynamics, electromagnetism, thermodynamics including statistical mechanics) probabilities are governed by underlying causal law. It is because of a lack of complete knowledge that probability arises; think of the coin example above or statistical mechanics where probability arises because we do not know some or all of the initial conditions of the myriads of atoms or molecules. Probability in quantum mechanics is quite different in that probability is an irreducible part of its interpretation. The mathematical formulation of the quantum theory, the matrix mechanics of Heisenberg in 1925 followed shortly after by the introduction of the Schrödinger equation in 1926, did not offer any obvious physical interpretation. Because, at that time, differential equations were so much more familiar than matrix methods, attention was focused on interpreting the solutions of the Schrödinger equation, called wave functions. The problem was that the wave function is complex and defied direct physical interpretation. Eventually, the interpretation of the wave function that came to be accepted was Max Born's probability interpretation. In other words, probability in quantum mechanics is fundamental and irreducible, and in many cases yields results that cannot be explained by classical probability. An alternative interpretation of quantum mechanics exists, called the causal interpretation (first suggested by L. de Broglie in 1926 in an incomplete form, and completed independently by D. Bohm in 1952) which offers a description of the quantum world in terms of classical pictures (but with crucial non-classical features) and describes, for example, an electron as a particle moving along some path. But, even in the causal interpretation, probability enters in a fundamental and irreducible way since initial positions remain hidden.

5.1.1 Interpretations of Probability

A number of interpretations of probability are possible, and there is a great deal of discussion among philosophers concerning which one is best, since whichever interpretation is adopted, a host of objections can be raised. Though this discourse is interesting, we do not need to consider all of the subtle nuances relating to it.

But, there is a crucial, inescapable feature that emerges from this philosophical discourse, namely, the dichotomy of two broad categories of interpretation. These are the *objectivist* or *objective* interpretation of probability and the *subjectivist* or *subjective* interpretation of probability.

Some workers favour one interpretation over the other, while others, the pluralists, take the view that the objective interpretation is more suited to some applications, while the subjective interpretation is more suited to others. I fall in the category of the pluralists for two reasons. First, because the areas to which probability is applied is diverse; compare the simple case of the probability of obtaining heads in the toss of a coin with the probability of deciding which horse will win a race. The first involves a few well defined factors, while the second not only involves many factors, but the factors themselves are not well defined. Deciding the winning horse depends on personal knowledge, belief and even prejudice. Second, even founding fathers of probability swayed from one interpretation to the other. Thus, adopting the simplest versions of both interpretations provides us with an intuitive working interpretation of probability which will be a great help in understanding the essential elements of probability and statistics.

Within the two categories there are numerous subtle and not so subtle variations. We will adopt and give the simplest definitions for each:

An *objective interpretation* refers to properties or behaviour that belong to a physical entity and is totally mind-independent (it does not depend on the knowledge, belief, prejudice or ignorance of any person). Objective probability is defined as the relative frequency of an event occurring (desired outcome) when very many trials are performed. Relative frequency is defined as the number of times the event of interest occurs divided by the number of trials. As an example, consider tossing a coin (we call each toss of the coin a trial) and ask for the probability of heads up. The probability of obtaining heads is found by tossing a coin, say, 1000 times. Suppose we get 506 heads. Then the probability of getting heads = relative frequency $= \frac{506}{1000} = 0.506$. We know from the many trials that have been performed that the probability of getting heads-up approaches 0.5, so our example gives a result close to this value. But, suppose we repeat the 1000 tosses and obtain 356 heads giving a probability of 0.356, far away from 0.5. Repeating the 1000 throws numerous times shows that the outcome of 0.365 is far less likely than that of getting a result close to 0.5. As we increase the number of trials, the chance of getting a result far from 0.5 becomes even more unlikely, Therefore, we can approach the true probability of 0.5 as close as we wish by taking a sufficient number of trials.[1] Thus, the simplest objective interpretation of probability is in terms of relative frequency. The objective interpretation is an a posteriori[2] interpretation of probability, since the probability is assigned after the trials are completed.

---

[1] A criticism of this definition is that however large the number of trials, a very skewed series of outcomes is still possible. Though, for a sufficiently large number of trials, the possibly of a skewed result becomes vanishingly small (such that, in the lifetime of the universe it never occurs), it is still possible. For all practical purpose this does not present a problem, but for a theoretical purist it remains an issue.

[2] Latin, meaning 'from what comes after'.

Objective Interpretation of Probability. (**An A Posteriori Interpretation**)
The probability of a desired outcome is defined as equal to the relative frequency
of the desired outcome occurring when very many trials are performed.

Relative Frequency is defined as the number of times the desired outcome
occurs divided by the total number of trials.

A trial (also called a random experiment) is an action or series of actions
leading to a number of possible outcomes.

A *subjective interpretation* refers to knowledge, belief, or even prejudice and is
therefore epistemic (meaning that it refers to human knowledge). Including 'belief'
and 'prejudice' provides a more general definition which is helpful, even necessary in
some applications (e.g. in estimating the probability that your favourite team will win
the World Cup). But, for our more scientific orientation we will restrict ourselves to a
definition in terms of the knowledge of a rational (able to think sensibly or logically)
person. Thus, according to a subjective interpretation, a person will assign a probability to an event according to the knowledge he or she possesses. As more knowledge
is gained, the probability assignment may change. The probability assignment may
differ from person to person depending on the knowledge they posses. Some subjectivists, but not all, maintain that rational persons possessing the same knowledge
will assign the same probability to the same event. Given our scientific focus, we
will adopt the latter assumption. All subjectivists agree that probability assignments
must satisfy the axioms of probability (as set out, for example, in the authoritative
1956 work of Kolmogorov, see the bibliography).

Now, the requirement that all rational minds with the same knowledge must assign
the same probability to an event requires a rule for assigning a numerical probability
value to an event in a way that does not allow for arbitrary differences of opinion.
Such a rule is the *Principle of Insufficient Reason* (PIR), which states that if there is no
reason to favour one or more of a total set of mutually exclusive outcomes (outcomes
where if one occurs the other cannot), each outcome is assigned the same probability.
Applications of this rule can be found throughout the early history of probability,
long before the principle was named. Since the probability is assigned before any
trial, the subjective interpretation is an a priori[3] interpretation of probability.

In the subjective interpretation, using PIR, we assign a probability of 0.5 of getting
heads when a coin is tossed, since there are only two mutually exclusive outcomes,
and since there is no reason to favour heads or tails for a fair coin. Clearly, the
probabilities in the coin tossing example (as is also the case for dice throwing, or
picking cards from a deck of cards) lends itself to either an objective or a subjective
interpretation.

A story given by Jacques Bernoulli of three ships setting sail from a harbour is
an example where a subjective interpretation is much more suited than an objective
interpretation. At some later time news arrives that one ship has sank. With no

---

[3]Latin, meaning 'from what is before'.

further information, PIR is applied and a probability of $\frac{1}{3}$ that a particular ship has sunk is assigned. But, suppose somebody knows that one ship is badly maintained or the captain is incompetent. With this extra knowledge, a rational person assigns a higher probability that that particular ship has sunk. There is no straightforward way, however, of assigning a numerical value to the increased probability.

Subjective Interpretation of Probability. (**An A Priori Interpretation**)
Probability is assigned according to the The Principle of Insufficient Reason (PIR).

Note 1 : Assigning probabilities according to PIR guarantees that rational persons possessing the same knowledge assign the same probability to the same event.

Note 2 : Probabilities assigned according to PIR satisfy the axioms of probability.

The Principle of Insufficient Reason states that if there is no reason to favour one or more of a total set of mutually exclusive outcomes, each outcome is assigned the same probability.

We repeat again that the definitions of objective and subjective interpretations we have given above are the simplest ones. To resolve problems with these definitions, more refined ones have been proposed. Since our objective here is to provide intuitive working definitions we do not need to consider these. We note, however, that PIR has been heavily criticised by numerous eminent critics. An example is Hans Reichenbach who espouses the common criticism that the definition of PIR is circular. Further, PIR is clearly not suited to applications where there are an infinite number of outcomes. Despite this, PIR provides an intuitive way of assigning probabilities and works very well in many applications. It is more than adequate for most of our purposes.

In the next section we give a brief history of the emergence of probability and statistics. We hope that our presentation, though necessarily selective, captures the main contributions.

5.1.2   A Brief History of the Emergence Probability and Statistics

Probability as we know it today was introduced around 1655–1660 and from the beginning was dual in having to do with degrees of knowledge (belief, even prejudice) on the one hand, and tendency (frequency) on the other. In other words, the dichotomy of the objective interpretation versus the subjective interpretation was present at the earliest origins of probability. Before 1655–1660 there was little formal development of probability, but some rough notions of probability and statistics existed from very early times. Sentences may be found in Aristotle's writing which translate as 'the probable is usually what happens'. Sextus Empiricus (A.D. 200) commented on 'signs' (e.g., natural signs like 'smoke' indicate fire, patients show 'signs' indicating disease). Signs, which may be natural or otherwise, were associated with probability and their use dates back to Aristotle. Gambling is ancient, possibly primeval. Dicing, for example, is one of the oldest human pastimes. A predecessor of dice is the astragalus or talus. The talus is the knuckle bone or heel of a running animal such as a deer, horse, ox, sheep or hartebeeste. The talus has the feature that when it is thrown it can come to rest in only four ways, hence its use for gaming. It is therefore surprising that an explicit formal development of probability did not take place much earlier. But, this does not mean that related, even detailed, statistical notions did not exist before 1655–1660. The idea of 'long observation', e.g., many throws of a dice, was known long before this period. It appears in the first book about games of chance written by Cardinano around 1550 but not published until 1663. Galileo Galilei (1564–1642) had a good sense of chance. He was also aware of the value of 'long observation', and even had a notion of relative frequency. The idea of experiment was emerging as an important tool of science, advocated by the philosopher Francis Bacon (1564–1642), and put into practice by Galileo. Galileo was perhaps the first to start taking averages of observations and he had a sophisticated appreciation of dealing with discrepant observations.

While gaming was the main initial driving force in the development of probability and statistics, economic and scientific interest proved a powerful incentive for further development. On the economic side, the need to calculate annuities (an annual payment from an insurance or investment) played an important role in driving the early development of probability and statistics and we will cite some of the main contributions below. On the scientific side, the need to calculate the macroscopic thermodynamic behaviour of gases in terms of the random movement of molecules or atoms drove the development of statistical mechanics. Again, we will cite the main contributions below.

In 1654 Pascal solved two problems of probability (one, to do with dice and one to do with dividing the stake among gamblers if the game is interrupted) and sent them to Fermat. The problems had been around for a long time and the chief clue to their solution, the arithmetic triangle, was known from about a century before. Despite this, Pascal's 1654 solutions are generally taken as the beginning of probability, sparking interest and setting the probability field rolling.

In 1657 Huygens wrote the first published probability textbook. It first came out in Latin in 1657 under the title *De ratiociniis in aleae ludo* (*Calculating in Games of Chance*). The Dutch version only appeared in 1660 as *Van Rekiningh in Spelen van Geluck*. Huygen's book is entirely about games of chance, with hardly any epistemic references. It is notable, if not ironic, that the word 'probability' does not appear in his book. About the same time Pascal made the first application of probability to problems other than games of chance, thereby inventing decision theory. Also at this time, the German law student Leibnitz, whilst still a teenager, applied probability ideas to legal problems. In 1665 Leibnitz proposed to measure degrees of proof and of right in law on a scale of 0 to 1, subject to a crude calculation of what he called probability. He also embarked on writing the first monograph on the theory of combinations.

Among the first contributions to the development of probability and statistics, motivated by the calculation of annuities were those in the late 1660s when John Hudde and John de Witt began determining annuities, based on the analysis of statistical data. A little earlier, in 1662, John Graunt published the first extensive set of statistical inferences drawn from mortality records.

The 1662 book Port Royal *Logic* was the first book to mention the numerical measurement of something actually called probability. There are conflicting reports on the authorship, but it is thought that Antoine Arnauld (1612–1694) was the main contributor, in particular writing Book IV which contains the chapters on probability. Arnauld was considered a brilliant theologian and philosopher. He was a member of the Jansenist[4] enclave at Port Royal. Pierre Nicole (1625–1695), who also contributed to *Logic*, and Blaise Pascal (1623–1662) were also members of the enclave. Pascal did not contribute to *Logic*.

The Principle of Insufficient Reason, under the name of 'equipossibility', originated with Leibnitz in 1678. He used it to define probability as the ratio of favourable cases to the total number of equally probable cases. Laplace, much later, also defined probability in this way. It is a definition that was prominent from the early beginnings of probability and is still in use today despite, as mentioned above, heavy criticism from numerous eminent critics.

Jacques Bernoulli's 1713 book *Ars conjectandi* (*The Art of Conjecturing*) represents a major contribution to the development of probability. The chief mathematical contribution was the proof of the *First Limit Theorem*. One of the founders of modern probability, A. N. Kolmogorov, commented that the proof was made with 'full analytical rigour'. The proof was given in 1692 but Bernoulli was not satisfied with it and the book was not published. Bernoulli died in 1705 and the book was passed on to the printer by his nephew Nicholas. It was published posthumously in 1713 in Basle.

J. Bernoulli was a subjectivist. Indeed, the version of the subjective interpretation we gave earlier largely originated with him. Though Bernoulli believed that a probability assignment to a given event may differ from person to person, depending on what information they possess, he also believed that rational persons in possession of the same information will assign the same probability to a given event; at least, there is no indication that he believed otherwise. In the latter view, he differed from modern subjectivists, like Bruno de Finetti, who interpret probability as being a measure of a rational degree of belief (which may depend on taste, prejudice, etc.). All subjectivists, however, require that probability assignments satisfy the axioms of probability.

J. Bernoulli recognised the difference between a priori and a posteriori methods of assigning probabilities. Just as Leibniz and Huygens used an early version of PIR, so also did Bernoulli, and he based the a prior method on it. But, Bernoulli recognised that PIR could not be applied completely generally and wanted to supplement it by the

---

[4]Jansenism is an austere form of Catholicism.

a posteriori method of frequency (as objectivists do). He claimed that by increasing the number of trials, the relative frequency would approach the probability with a certainty as close to 1 as one wishes. To support this claim, he proved what is now called the (*Weak*) *Law of Large Numbers*. For subtle reasons we will not go into, his Law of Large Numbers failed to justify his claim.

Reverend Thomas Bayes addressed Bernoulli's problem in a famous essay published posthumously in 1763. Bayes' was an ardent subjectivist and modern subjectivists sometimes call themselves Baysians. In Bayes formulation of the problem he asks about the probability of a probability given certain data. Bayes succeeded in providing a solution but only at the price of introducing the new notion of probabilities of probabilities. This raised the new problem of justifying such an assignment. Bayes saw his solution as being an aspect of PIR, and that PIR was inescapable, very much in line with his ardent subjectivist views.

Modern subjectivists after the 1920s, such as F. P. Ramsey, B. de Finetti and, somewhat later, L. Savage and R. C. Jeffrey adopt the view that rational minds possessing the same information need not agree on probability assignments, since they consider that probability assignments depend on a persons system of beliefs which may include anything from expert opinion to sheer prejudice. Thus, they do not need a principle such as PIR.

Another notable contribution motivated by annuities was the production in 1693 of the best mortality tables of the time by Astronomer Royal Edmond Halley with help from Leibnitz. These tables remained a standard for about eighty years. De Moivre's classic 1725 textbook on annuities included Halley's tables, which were the basis of the computations by de Moivre. De Moivre was regarded as one of the finest probabilists of his time.

The first notable contribution to statistical physics, specifically to kinetic theory of gases, was by Daniel Bernoulli, the nephew of Jacques Bernoulli, in his 1738 work *Hydrodynamica*. He considered a gas as composed of molecules (or atoms) moving randomly. He explained pressure in terms of the momentum exchange when the molecules collide with the walls of the container and was also able to derive the *ideal gas law*. It is notable, even surprising considering he was the nephew of Jakob Bernoulli, that Daniel Bernoulli did not consider that probability was at all relevant in his analysis.

In 1812, Laplace published his authoritative book *Théorie analytique des probabilités* (*Analytic Theory of Probability*). In the second edition, published in 1814, he included an essay for the popular reader *Essai philosophique sur les probabilités* (*A Philosophical Essay on Probability*) as its introduction. In his book, he described the mathematical tools he invented for mathematically determining probabilities. Laplace was a subjectivist, much like J. Bernoulli, and shared the view that rational persons in possession of the same knowledge assigned similar probabilities to events, as well as using PIR to assign probabilities. This view of probability, commonly held from J. Bernoulli to Laplace, is often called 'classical'.

More than a century after D. Bernoulli, Maxwell added his significant contribution to the development of statistical mechanics. Maxwell was well aware of the classical theory of probability originated by J. Bernoulli, and started out, like him, as a subjectivist. But in his work on gases he adopted an objective interpretation of probability. This is evident in his definition of the probability that the velocity of a molecule lies in a certain interval. He defines this probability as the relative frequency of the molecules having velocities in this interval. This definition refers to objective fact without appeal to rational minds and so, clearly, is an objective interpretation of probability. It is thought that Maxwell's change of view was influenced by frequentists (proponents of the objective frequency interpretation) such as Leslie Ellis (like Maxwell, a fellow of trinity college, Cambridge), George Boole and John Venn.

In his famous 1860 derivation of what is now called the *Maxwell Distribution Law* he introduced the velocity distribution function $f(v)$ which he assumes to be spherically symmetric, i.e., it depends only on the magnitude of the velocity vector, that is to say, on $v = |\vec{v}|$, and further assumes that it factorises into functions of the three orthogonal velocity components,

$$f(v) = \phi(v_x)\phi(v_y)\phi(v_z).$$

But, this assumption is in total opposition to a frequency interpretation of the distribution function $f(v)$, but rather, is justified by a version of PIR, namely, if there is no reason to suppose a connection between quantities, we are entitled to regard them as independent. But, this is a subjective interpretation.

In 1867 Maxwell presented a more detailed and much improved derivation of his distribution law, but the subjective element in the derivation remained. Maxwell and his reviewers were concerned with this dichotomy, but we take the view, that this exemplifies that both the objective and subjective interpretations of probability are needed to comprehend all the diverse applications of probability.

A little after Maxwell came the profound contributions to statistical mechanics of Ludwig Eduard Boltzmann. Boltzmann was an ardent objectivist who seems to have derived his knowledge of probability from Maxwell, mixing Maxwell's ideas with his own. However, he had less patience than Maxwell with the classical subjective interpretation. Though for Boltzmann probability was an objective quantity, five notions of probability can be identified in his works, some of them, subjective interpretations (But, we won't list them here as the differences between the notions are subtle and technical. Here, we only want to point out that these differences exist).

In 1927 Edwin Jaynes developed an extensive subjective interpretation of statistical mechanics. Jaynes interpretation is more controversial, even among fellow subjectivists. Though controversial, he did not go as far as to adopt the personalised subjective interpretation of Ramsey and de Finetti described earlier, but followed J. Bernoulli and Laplace by requiring that rational persons with the same information will assign the same probabilities. A notable part of his work was his *Maximum-Entropy Principle*, which is a generalisation of PIR.

In later developments of statistical mechanics, the notion of ergodic systems evolved. Ergodic systems are systems which in a long enough time pass through all points of phase space. A detailed specification for a system to be ergodic was given in the *Birkhoff-von Neumann Ergodic Theorem*. Generalisations of this theorem are considered important for an objective interpretation, since for ergodic systems time-averages can be equated to ensemble-averages. The reason this is important is that time-averages are objective and thus favoured over subjective ensembles, but time-averages are notoriously difficult to calculate, whereas calculations with ensembles are much easier. Thus, an objectivist would like to develop a probability interpretation in terms of time-averages, but use ensembles only for calculation, hence the importance of the equivalence of these averages. This is the case for D. A. Lavis who has refined and defends an objective interpretation of statistical mechanics. His basic idea is precisely to identify probabilities with time-averages. When a system is ergodic time-averages are well defined and probabilities can be easily defined. When systems are not ergodic the definition is more challenging. Lavis uses ergodic decomposition and other more recent notions to define probabilities for this case. In either case, the definitions are mind-independent.

It is appropriate, acknowledging numerous omissions, that we conclude with Kolmogorov,(Andrey Nikolayevich Kolmogorov, 1903–1987), a brilliant Russian mathematician and a founding father of modern probability. His motivation for the axiomatic approach is that despite the practical value of both the objective and subjective interpretations of probability, attempts to place these interpretations on a solid theoretical footing proved to be very difficult. In an early paper, *General Theory of Measure and Probability Theory*, he aimed to develop a rigorous axiomatic foundation for probability. He expanded this paper into the very influential 1933 monograph *Grundbegriffe der Wahrscheinlichkeitsrechnung*. It was translated into English and published in 1950 as *Foundations of the Theory of Probability*. He also made profound contributions to stochastic processes, especially Markov processes.

5.2 Basics

In the next section we begin our exposition of the essential elements of probability and statistics. In the rest of this chapter, unless otherwise stated, the outcomes of trials will be assumed equally probable (by the Principle of Insufficient Reason).

We begin with the definition of a trial, also called a random experiment. We recall again its definition:

> A **trial** (also called a **random experiment**) is an action or series of actions leading to a number of possible outcomes).

Example 5.2.1 (Throwing a dice)
The action of throwing a dice is an example of a trial. The outcome is not known; it can be any number between 1 to 6. In such a trial, we assume the die to be fair. It should be symmetrical and made of a homogeneous material so that no face is favoured.

Another important concept is the *sample space*. It is defined as follows:

> A **sample space** consistes of all possible outcomes of a trial. Each outcome is called a **sample point**.

Often, the outcomes of trials may be described by more than one sample space.

Example 5.2.2 (More than one sample space for the same trial)
Give two examples of a trial with more than one sample space and define at least two different sample spaces for each example.
Solution
In throwing a dice, the outcomes {1,2,3,4,5,6} form one possible sample space. The outcomes {even, odd} form another possible sample space. The outcomes from picking a card from a deck of cards can be represented by three different sample spaces: {52 cards}, {red, black} and {hearts, spades, clubs, diamonds}.

The sample space can be finite and hence called a *finite sample space*, or it can be infinite. If it is infinite, it may be countable (having as many points as natural numbers) and called a *countable infinite sample space*, or it may be noncountable as is the case for a number n defined in some interval $0 \leq n \leq 1$ and called a *noncountable infinite sample space*. Finite or countable sample spaces are also called *discrete sample spaces*, while a noncountable infinite sample space is also called a *nondiscrete or continuous sample space*. The sample spaces consisting of the outcomes of throwing a dice are discrete, while the sample space consisting of heights of students in a class is continuous. Detection counts of radioactive particles are an example of an infinite discrete sample space. The counts are discrete, but the full distribution is given by counting for an infinite length of time.

A group of sample points is called an *event*, while a single sample point is called a *simple event*. An example of an event is getting an odd number {1,3,5} when a dice is thrown. Simple events are {1}, {2}...{6}. When an event consists of all sample points, it is certain to occur; if it consists of none of the sample points, it cannot occur.

> An **event** is a group of sample points.
> A single sample point is called a **simple event**.

We are now ready to consider the mathematical elements of probability. Consider first the objective interpretation or a posteriori interpretation. We perform a large number n of trials and suppose that event A occurs $f(A)$ times. The number of times $f(A)$ that A occurs is called the *frequency* of A occurring, while the ratio $f_r(A) = \frac{f(A)}{n}$ is called the *relative frequency* that A occurs. Then, as earlier, the objective

interpretation of probability is defined as equal to the relative frequency for a large
number of trials n:

Definition of Objective Probability
$P(A) = f_r(A) = \dfrac{f(A)}{n} = \dfrac{\text{number of times } A \text{ occurs}}{\text{number of trials}}$
$f(A)$ = frequency of A
$f_r(A)$ = relative frequency of A

As we mentioned in Sect. 5.1.2, that the relative frequency $\frac{f(A)}{n}$ actually approaches
the probability with a certainty arbitrarily close to 1 as the number of trials is increased
is difficult to prove rigorously, as we saw with J. Bernoulli's attempt to prove this limit.
Practically, however, this definition is found to work extremely well. For example,
when a coin is tossed 1000 times, say, we do find that the relative frequency $f_r(A)$ of
getting heads-up is indeed very close to 0.5. We note that the probability $P(A)$ must
be a number between 0 and 1.

In the subjective interpretation we assign probabilities according to the Principle
of Insufficient Reason defined in Sect. 5.1.1. In this interpretation n is interpreted as
the total number of mutually exclusive, equally probable outcomes of a trial, while
$f(A)$ is the number of ways A can occur. Then, the probability $P(A)$ of getting A is
$P(A) = \frac{f(A)}{n}$.

Definition of Subjective Probability
$P(A) = \dfrac{f(A)}{n} = \dfrac{\text{number of ways } A \text{ can occurs}}{\text{total number of outcomes}}$

Mutually exclusive events are events for which the occurrence of one precludes the occurrence of the other.

We note that both formula look the same, but we should keep in mind that they differ
in interpretation.

Example 5.2.3 (Calculation of probabilities I)
Consider tossing a coin. State how to establish the probability of getting a head using
the subjective (PIR) and objective interpretations of probability.
Solution. There are a total of two possible outcomes and one desired outcome, heads-
up. By PIR, the two possible outcomes are equally probable, so that the probability
of getting a head is a $\frac{1}{2}$. Using the objective interpretation, we would either first have
to perform many trials ourselves or else use pre-existing results established by the

performance of many trials to establish that the probability of getting a head should be $\frac{1}{2}$.

Example 5.2.4 (Calculation of probabilities II)
What is the probability of getting an ace when drawing a card from a deck of 52 cards?
Solution
We first note that by PIR all cards have an equal probability of being drawn, and since there are 52 cards and four desired outcomes (drawing an ace) this probability is $\frac{4}{52}$. We note that in this kind of probability calculation the subjective interpretation is clearly better suited.

In many of the examples that follow we will use simple trials such as tossing a coin, throwing a dice, drawing a card from a deck of 52 cards, or choosing coloured balls from a bag. For such simple cases, it is convenient to use PIR to assign equal probabilities to the outcomes. After noting this, in the remainder of this chapter, we will not continue to specify which interpretation we are using. We will assume that coins, dice etc. are fair. A dice should be symmetrical and homogeneous, a deck of cards should not contain 6 aces, a coin should not have two heads and so on. We assume all of this in the examples that follow in the remainder of this chapter.

5.2.1 The Axioms of Probability

Though we have only briefly alluded to difficulties with both the objective and subjective interpretations, mainly because the criticisms are technically advanced, the difficulties are viewed as severe. For this reason, modern probability is based on an axiomatic approach pioneered by Kolmogorov, and we list the basic axioms below. The objective and subjective interpretations are still valuable since they provide a more intuitive understanding of probability and we will keep these interpretations in mind in what follows. Before stating the axioms, we first introduce some commonly used notation:

$\cup$ = union	$\cap$ = intersection	$\subset$ = subset.

Let A and B be two events belonging to a sample space $\mathcal{S}$:

$P(A \cup B)$ = Probability that A **or** B occurs
$P(A \cap B)$ = Probability that both A **and** B occur
$A \subset \mathcal{S} = A$ is a subset of $\mathcal{S}$.

Some more needed notation:

$P(A') = $ Probability that A doesn't occur
$P(A - B) = $ Probability that A occurs but B doesn't

Let S be a sample space, which may be discrete or continuous. For a discrete sample space all subsets can be taken to be events, but for a continuous sample space only subsets satisfying certain mathematical conditions (which are too technical for us to state) can be considered as subsets. We saw earlier that the outcomes of a particular trial can be represented by different sample spaces (see Example 5.2.2). We therefore need to differentiate sets of events belonging to different sample spaces, and we do this by placing the sets of events into classes. Thus, we will consider event A of a given trial and sample space as belonging to class C, and allocate to it a number $P(A)$ between 0 and 1. Then $P(A)$ can be interpreted as the probability of the event occurring if the following axioms are satisfied:

Axiom 1	For every event A in class C $$0 \le P(A) \le 1$$
Axiom 2	For an event S in class C which is certain to occur $$P(S) = 1$$

Axiom 3 states that for mutually exclusive events $A_1, A_2 \ldots$ in class C, the probability of occurrence of any one of them is equal to the sum of the probabilities of each of them occurring:

Axiom 3	$P(A_1 \cup A_2 \cup \ldots) = P(A_1) + P(A_2) + \cdots$

For only two mutually exclusive events, axiom 3 reduces to

$$P(A_1 \cup A_2) = P(A_1) + P(A_2).$$

When these three axioms are satisfied $P(A)$ is called a probability function. There are a number of theorems which we need to note.

Theorem 5.2.1 *Consider an event A_2 made up of a number of occurrences of a given trial. Let event A_1 be a subset of A_2, then, perhaps rather obviously, the probability $P(A_1)$ that A_1 occurs is less than the probability $P(A_2)$ that A_2 occurs. Also, the probability $P(A_2 - A_1)$ that A_2 occurs but A_1 doesn't occur is the difference of the probabilities of each of them occurring:*

Theorem 5.2.1	If $A_1 \subset A_2$ then $P(A_1) \leq P(A_2)$ and $P(A_2 - A_1) = P(A_2) - P(A_1)$

Example 5.2.5 (An example of the use of Theorem 5.2.1)
Let event A_2 consist of the diamond suite of a deck of 52 cards, and let event A_1 consist of the odd-numbered diamonds (we take the jack=11 and the king=13). The trial consists of picking a card from the deck. Show that $P(A_2) \geq P(A_1)$ and determine $P(A_2) - P(A_1)$ by Theorem 5.2.1. Confirm your answer for $P(A_2) - P(A_1)$ by direct counting.
Solution
By counting desired outcomes, the probabilities are $P(A_1) = \frac{7}{52}$ and $P(A_2) = \frac{13}{52}$. Then, clearly, $P(A_2) \geq P(A_1)$. Also $P(A_2) - P(A_1) = \frac{6}{52}$. By direct counting of desired outcomes, the number of ways of getting A_2 but not A_1 is 6, giving a probability $P(A_2) - P(A_1)$ of getting A_2 but not A_1 of $\frac{6}{52}$, confirming the result from Theorem 5.2.1.

Theorem 5.2.2 *The probability of not getting an outcome from the trial is 0. Let $\emptyset$ represent no event (or empty set), then*

Theorem 5.2.2	$P(\emptyset) = 0$

Theorem 5.2.3 *The probability $P(A')$ of an event A not occurring is equal to 1 minus the probability $P(A)$ of A occurring. This follows since, as is perhaps obvious, the probability of A occurring plus the probability of A not occurring covers all possible outcomes, i.e., $P(A) + P(A') = 1$:*

Theorem 5.2.3	$P(A') = 1 - P(A)$

Theorem 5.2.4 *Let A be an event made up of n mutually exclusive events $A_1, A_2 \ldots A_n$, i.e., $A = A_1 \cup A_2 \ldots \cup A_n$, then*

Theorem 5.2.4	If $A = A_1 \cup A_2 \ldots \cup A_n$, $A_1, A_2, \ldots A_n$ mutually exclusive, then $P(A) = P(A_1) + P(A_2) + \cdots + P(A_n)$

For the special case A = S

Theorem 5.2.4	*If* $A = A_1 \cup A_2 \ldots \cup A_n$, $A_1, A_2, \ldots A_n$ *mutually exclusive, then* $P(S) = P(A_1) + P(A_2) + \cdots + P(A_n) = 1$

Example 5.2.6 (An example of the use of Theorem 5.2.4)
Let A be composed of the events $A_1 = 4$ *of hearts*, $A_2 = 4$ *of diamonds*, $A_3 = 4$ *of clubs* and $A_4 = 4$ *of spades*. The trial consists of drawing one card from a deck of 52 cards. The event A of interest is drawing one of these cards. Find $P(A)$ both by direct counting of desired outcomes and by Theorem 5.2.4, and hence, confirm 5.2.4.
Solution
By counting desired outcomes in each case we find the probabilities to be $P(A) = \frac{4}{52}$, $P(A_1) = \frac{1}{52}, P(A_2) = \frac{1}{52}, P(A_3) = \frac{1}{52}$ and $P(A_4) = \frac{1}{52}$. Then

$$P(A) = P(A_1) + P(A_2) + P(A_3) + P(A_4) = \frac{1}{52} + \frac{1}{52} + \frac{1}{52} + \frac{1}{52} = \frac{4}{52},$$

which confirms Theorem 5.2.4.

Theorem 5.2.5 *For any two events A or B, not necessarily exclusive, the probability* $P(A \cup B)$ *is given by*

Theorem 5.2.5	$P(A \cup B) = P(A) + P(B) - P(A \cap B)$

Example 5.2.7 (An example of the use of Theorem 5.2.5)
Let A = *hearts suite of a deck of* 52 *cards* and B = *ace plus cards numbered from 2 to* 6. The trial is drawing a card. Use Theorem 5.2.5 to find the probability $P(A \cup B)$ of drawing a heart or a card numbered 1 to 6. Confirm your answer by counting desired outcomes.
Solution
The probability of choosing a heart is $P(A) = \frac{13}{52}$, while the probability of drawing a card numbered 1 to 6 is $P(B) = \frac{24}{52}$. The probability of drawing a heart numbered 1 to 6 is $P(A \cap B) = \frac{6}{52}$. Then

$$P(A \cup B) = P(A) + P(B) - P(A \cap B) = \frac{13}{52} + \frac{24}{52} - \frac{6}{52} = \frac{31}{52}.$$

The probability of drawing a heart or a card from ace to 6 by counting desired outcomes is $P(A \cup B) = \frac{31}{52}$, where hearts ace to 6 are only counted once, confirming the answer using Theorem 5.2.5.

The generalisation of Theorem 5.2.5 to n events is possible, but cumbersome. To give an idea of the generalisation we write Theorem 5.2.5 for the case of three events

$$P(A_1 \cup A_2 \cup A_3) = P(A_1) + P(A_2) + P(A_3) - P(A_1 \cap A_2) - P(A_1 \cap A_3)$$
$$-P(A_2 \cap A_3) + P(A_1 \cap A_2 \cap A_3).$$

Theorem 5.2.6 *For any two events, mutually exclusive or not, the probability $P(A \cap B)$ that A and B occur together plus the probability $P(A \cap B')$ that A and any element not in B occur together is equal to the probability that A occurs:*

Theorem 5.2.6	$P(A) = P(A \cap B) + P(A \cap B')$

Example 5.2.8 (An example of the use of Theorem 5.2.6)
Consider again events A and B and the trial of example 5.2.7. By calculating the three probabilities $P(A)$, $P(A \cap B)$ and $P(A \cap B')$ by direct counting, confirm theorem 5.2.6.
Solution
The probabilities, by direct counting are $P(A) = \frac{13}{52}$, $P(A \cap B) = \frac{6}{52}$ and $P(A \cap B') = \frac{7}{52}$, since event B' consists of all numbers from 7 to 13. Substituting into Theorem 5.2.6 we get

$$P(A) = P(A \cap B) + P(A \cap B') = \frac{6}{52} + \frac{7}{52} = \frac{13}{52},$$

in agreement with Theorem 5.2.6.

Theorem 5.2.7 *If an event A consists of mutually exclusive events $A_1, A_2, \ldots A_n$ then the probability that A occurs is equal to the sum of probabilities that A occurs with each $A_1, A_2, \ldots A_n$:*

Theorem 5.2.7	*If $A = \{A_1, A_2, \ldots A_n\}$, $A_1, A_2, \ldots A_n$ mutually exclusive, then* $P(A) = P(A \cap A_1) + P(A \cap A_2) + \ldots + P(A \cap A_n)$

Example 5.2.9 (An example of the use of Theorem 5.2.7)
Let A = *the diamond suite of a deck of* 52 *cards*. It consists of the mutually exclusive events: A_1 = *diamonds ace to* 5, A_2 = *diamonds* 6 *to* 9 and A_3 = 10, *jack, queen, king of diamonds*. The trial consists of drawing a card. By determining the probability $P(A)$ that a diamond is drawn and the probabilities of A occurring with A_1, A_2, and A_3, confirm Theorem 5.2.7.
Solution
By counting desired outcomes we get $P(A) = \frac{13}{52}$, $P(A_1 \cap A) = \frac{5}{52}$, $P(A_2 \cap A) = \frac{4}{52}$ and $P(A_3 \cap A) = \frac{4}{52}$, then

$$P(A) = P(A \cap A_1) + P(A \cap A_2) + P(A \cap A_3) = \frac{5}{52} + \frac{4}{52} + \frac{4}{52} = \frac{13}{52}$$

in agreement with Theorem 5.2.7.

5.2.2   Conditional Probability

The notation $P(B|A)$ denotes the probability that B occurs given that A has already occurred.

$P(B|A)$ = probability that B occurs given that A has already occured

Consider two events A and B which in general will not be mutually exclusive. Further suppose that A and B belong to a sample space $\mathcal{P}$ consisting of events $\{A, B, C, \ldots\}$ with M total events (noting that each event can appear more than once). Suppose that event A occurs m times in $\mathcal{P}$, so that $P(A) = \frac{m}{M}$. We consider two trials. The first results in one event of the sample space $\mathcal{P}$ occurring, and the second also results in one of the events of $\mathcal{P}$ occurring. This defines a sample space consisting of all combinations of events $A, B, C, \ldots$ in pairs. We call this the original sample space and denote it by S. We suppose S has N sample points (pairs of events). The occurrence of event A defines a new reduced sample space (a subset of the original sample space S). The new sample space is made of all the pairs of events of the original space in which A occurs first. We denote the new sample space by S_r and suppose it has n sample points. Let h be the number of times A and B both occur in S_r, so that $P(A \cap B) = \frac{h}{N}$. We define the probability $P(B|A)$ with respect to the new sample S_r space as follows:

$$P(B|A) = \frac{\text{number of times } A \text{ and } B \text{ both occur in } S_r}{\text{total number of sample points of } S_r} = \frac{h}{n}$$

Now divide the numerator and denominator by N to get an important formula for conditional probabilities

$$P(B|A) = \frac{h/N}{n/N} = \frac{h/N}{m/M} = \frac{P(A \cap B)}{P(A)}$$

We have not proved the that $\frac{n}{N} = \frac{m}{M}$, but state that it is necessarily true and justify our claim by the example to follow. Hopefully, this example will also help to clarify our argument leading to the important formula above, which we highlight as:

$$\boxed{\text{Formula for Conditional Probability}} \quad \boxed{P(B|A) = \frac{P(A \cap B)}{P(A)}} \quad (5.1)$$

Rearranging formula (5.1) leads to the important multiplication rule

$$\boxed{\text{Multiplication Rule}} \quad \boxed{P(A \cap B) = P(B|A)P(A)} \quad (5.2)$$

Example 5.2.10 (Conditional probability. Sampling with replacement)
Consider a bag containing 2 black balls and 3 red balls. The trial is picking a ball and is repeated twice. Determine the probability of picking a red ball, given that a red ball is already picked. Trials consisting of choosing or picking objects are actions referred to as sampling. Sampling can take place in two ways: (a) sampling with replacement, and (b) sampling without replacement. We will consider (a) in this example and (b) in a later example.
Solution
Let event $A =$ *first event, red ball picked*, and event $B =$ *second event, red ball picked*. Noting that $M = 5$ and $m = 3$, the probability $P(A) = \frac{m}{M} = \frac{3}{5}$. The probability we require is the conditional probability $P(B|A)$. We will calculate this probability by first setting up the original and new reduced sample spaces, then, count the total elements of the each sample space and count the desired outcomes. The original

sample space consists of all pairs of balls with the first ball replacement after being picked:

	RR	RR	RR	RB	RB
Original	RR	RR	RR	RB	RB
sample	RR	RR	RR	RB	RB
space	BR	BR	BR	BB	BB
	BR	BR	BR	BB	BB

The new reduced sample space consists of all pairs of balls in which a red ball is picked first:

New	RR	RR	RR	RB	RB
reduced	RR	RR	RR	RB	RB
sample space	RR	RR	RR	RB	RB

$P(B|A)$ is found using the new sample space. The number of occurrences of both A and B (i.e., the pairs RR) in the new sample space is 9. The total number of sample points in the new sample space is 15 so that the required probability $P(B|A)$ of picking a red ball given that one has already been picked is

$$P(B|A) = \frac{9}{15} \qquad (5.3)$$

To illustrate our argument leading to formula (5.1), and to offer some justification for equating $\frac{n}{N}$ to $\frac{m}{M}$, we first calculate $P(A \cap B)$ using the original sample space, which has 25 elements. In the original sample space A and B both occur (i.e., RR) 9 times, so that $P(A \cap B) = \frac{9}{25}$. Next, dividing the numerator and denominator of Eq. (5.3) by 25, we get

$$P(B|A) = \frac{9/25}{15/25} = \frac{9/25}{3/5} = \frac{P(A \cap B)}{P(A)}.$$

We see that $\frac{15}{25} = \frac{n}{N} = \frac{3}{5} = \frac{m}{M}$.

The conditional probability $P(B|A)$ is more easily calculated by counting desired outcomes. When the red ball is replaced there are once again 3 red balls and 2 black balls in the bag, so that the probability of picking a second red ball is simply $P(B) = \frac{3}{5} = P(B|A)$, in agreement with our above result. This result follows because the events A and B are statistically independent (see the next section). We may conclude that for sampling with replacement, events are statistically independent.

Example 5.2.11 (Multiplication rule I. Sampling with replacement)
We consider the ball-picking trials of example 5.2.10. Calculate the probability $P(A \cap B)$ of picking two red balls, with replacement, using the multiplication rule Eq. (5.2).

Solution

From Example 5.2.10 or by counting desired outcomes, $P(B|A) = \frac{3}{5}$ and $P(A) = \frac{3}{5}$, so that

$$P(A \cap B) = P(B|A)P(A) = \frac{3}{5} \cdot \frac{3}{5} = \frac{9}{25},$$

in agreement with the value calculated in Example 5.2.10 by counting desired outcomes in the reduced sample space.

Example 5.2.12 (Conditional probability. Sampling without replacement)
Consider again Example 5.2.10, but this time with the first picked ball not replaced. Again calculate the probability $P(B|A)$.

Solutions

Since the picked ball is not replaced both the original sample space and the new sample space of Example 5.2.10 are further reduced by removing the occurrences where the picked ball is paired with itself. The sample spaces for the case of no replacement becomes:

	RR RR RB RB
Original	*RR RR RB RB*
sample	*RR RR RB RB*
space	*BR BR BB BB*
	BR BR BB BB

New	*RR RR RB RB*
reduced	*RR RR RB RB*
sample	*RR RR RB RB*
space	

By counting the number of times A and B both occur (RR occurrences) in the reduced sample space, and noting that the number of sample points in the reduced sample space is 12, we immediately get

$$P(B|A) = \frac{6}{12} = \frac{1}{2}.$$

We can also get the same result more simply by directly counting the balls, since once a red ball is picked, 2 black balls and 2 red balls remain in the bag, giving

$$P(B|A) = \frac{2}{4} = \frac{1}{2},$$

in agreement with the above.

Example 5.2.13 (Multiplication rule II. Sampling without replacement)
Again, we consider Example 5.2.10. Find the probability $P(A \cap B)$ of picking 2 red balls from the bag using the multiplication rule Eq. (5.2), but, this time, without replacing the ball.

Solution
From example 5.2.12, or by counting desired outcomes, $P(B|A) = \frac{2}{4}$ and $P(A) = \frac{3}{5}$, so that

$$P(A \cap B) = P(B|A)P(A) = \frac{3}{5} \cdot \frac{2}{4} = \frac{6}{20} = \frac{3}{10}.$$

We can check this result by considering the original sample space of Example 5.2.12, which has 20 sample points. Since A and B both occur (RR occurrences) 6 times we immediately get

$$P(A \cap B) = \frac{6}{20} = \frac{3}{10},$$

confirming the above result.

As may have been noticed from the above examples, $P(A)$ and $P(B|A)$ are easily obtained since they can be calculated directly from the sample spaces of single trials, while calculating $P(A \cap B)$ is more tedious because it requires a consideration of the sample space of pairs of outcomes from two trials. Therefore, the conditional probability formula in the form of the multiplication rule Eq. (5.2) allows $P(A \cap B)$ to be calculated more easily.

Example 5.2.14 (Multiplication Rule III. Drawing cards I)
What is the probability of drawing a queen and a king from a deck of 52 cards, with and without replacement?

Solution
The probability $P(A \cap B)$ is the same whichever card is picked first. In this case, we suppose that the queen is picked first.

With replacement: Let A = *drawing a queen* and B = *drawing a king*. Since there are four queens in a deck of 52 cards $P(A) = \frac{4}{52}$ of picking a queen first. Similarly, after replacing the queen, the probability of drawing a king given that a queen is drawn first is $P(B|A) = \frac{4}{52}$. By the multiplication rule, the probability $P(A \cap B)$ of choosing a queen followed by a king is

$$P(A \cap B) = P(B|A)P(A) = \frac{4}{52} \cdot \frac{4}{52} = \frac{1}{169} = 0.0059.$$

Without replacement: Again, $P(A) = \frac{4}{52}$. But, since the queen is not replaced there are only 51 cards left in the card deck, so that $P(B|A) = \frac{4}{51}$, giving

$$P(A \cap B) = P(B|A)P(A) = \frac{4}{51} \cdot \frac{4}{52} = \frac{4}{663} = 0.0060.$$

We see that the probability is higher without replacement. This is expected since the probability of picking a king is slightly higher.

The following theorem is useful:

Theorem 5.2.8 *Generalisation of the multiplication rule.*

Theorem 5.2.8	The probability that any three events A_1, A_2, A_3 occur together is $P(A_1 \cap A_2 \cap A_3) = P(A_1)P(A_2|A_1)P(A_3|A_1 \cap A_2)$

It is not difficult to generalise this to n events.

Example 5.2.15 (Multiplication Rule IV. Drawing cards II)
What is the probability of drawing four aces from a deck of 52 cards, with and without replacement?

Solution
With replacement: Let the four events A_1, A_2, A_3, A_4 each represent drawing an ace. We need to use the multiplication rule generalised to four events

$$P(A_1 \cap A_2 \cap A_3 \cap A_4) = P(A_1)P(A_2|A_1)P(A_3|A_1 \cap A_2)P(A_4|A_1 \cap A_2 \cap A_3)$$

Since the aces are replaced all four probabilities are the same, i.e., $P(A_1) = P(A_2|A_1) = P(A_3|A_1 \cap A_2) = P(A_4|A_1 \cap A_2 \cap A_3) = \frac{4}{52} = \frac{1}{13}$. Then the probability $P(A_1 \cap A_2 \cap A_3 \cap A_4)$ of drawing four aces is

$$P(A_1 \cap A_2 \cap A_3 \cap A_4) = \frac{1}{13} \cdot \frac{1}{13} \cdot \frac{1}{13} \cdot \frac{1}{13} = \frac{1}{28,561} \approx 3.5 \times 10^{-5}.$$

Thus, a poker player's dream hand is not going to be realised very often.
Without replacement: This time, the conditional probabilities are different. As before $P(A_1) = \frac{4}{52}$, but since the first ace is not replaced only 3 aces and 51 cards remain so that $P(A_2|A_1) = \frac{3}{51}$. Similarly $P(A_3|A_1 \cap A_2) = \frac{2}{50}$ and $P(A_4|A_1 \cap A_2 \cap A_3) = \frac{1}{49}$. The probability of drawing four aces without replacement is thus given by

$$P(A_1 \cap A_2 \cap A_3 \cap A_4) = \frac{4}{52} \cdot \frac{3}{51} \cdot \frac{2}{50} \cdot \frac{1}{49} = \frac{1}{270,725} = 3.7 \times 10^{-6}.$$

As we might have guessed, the probability is very much smaller without replacement.

Here is another useful theorem:

Theorem 5.2.9

Theorem 5.2.9	*If an event A consists entirely of events $A_1, A_2, \ldots A_n$ then* $P(A) = P(A_1)P(A\|A_1) + P(A_2)P(A\|A_2) + \ldots + P(A_n)P(A\|A_n)$

5.2.3   Independent Events

We saw above that for sampling with replacement

$$P(B|A) = P(B),$$

which means that the probability of B is not affected by the probability of A. In this case A and B are said to be *independent events*. For this important special case of independent events the multiplication rule reduces to:

Independent Events	$P(A \cap B) = P(A)P(B)$	(5.4)

Conversely, events A and B can be recognised as being independent whenever Eq. (5.4) holds.

Similarly, for three independent events A_1, A_2, A_3,

$$P(A_1 \cap A_2 \cap A_3) = P(A_1)P(A_2)P(A_3). \tag{5.5}$$

For three events to be independent, they must be pairwise independent:

$$P(A_i \cap A_j) = P(A_i)P(A_j), \quad i \neq j, \quad i = j = 1, 2, 3.$$

Equation (5.5) is easily generalised to n independent events

$$P(A_1 \cap A_2, \cap A_3, \ldots, \cap A_n) = P(A_1)P(A_2)P(A_3)\ldots P(A_n). \tag{5.6}$$

A useful formula for more advanced applications is Bayes' rule, also called Bayes' theorem: Let $A_1, A_2, \ldots, A_n$ be mutually exclusive events covering the whole sample space, so that one of these events must occur. Let A be another event of the same sample space. Then, Bayes' theorem states:

$$P(A_i|A) = \frac{P(A_i)P(A|A_i)}{\sum_{i=1}^{n} P(A_i)P(A|A_i)}.$$

It is perhaps easier to see the meaning of the theorem after rearranging it:

$$P(A|A_i) = \frac{P(A_i|A) \sum_{i=1}^{n} P(A_i)P(A|A_i)}{P(A_i)}$$

This gives us the probability of A occurring given that A_i has occurred. That one of the events A_i must occur, means that A can only occur if one of the A_i's has occurred. Hence, in a sense, events A_i can be thought of as causing A to occur. For this reason Bayes' theorem is sometimes viewed as a theorem on probable causes.

5.2.4 Permutations

In simple cases probabilities can be determined by counting desired outcomes. For more complicated situations, methods of counting different ways of arranging objects are helpful in determining probabilities. These methods of counting come under the name of *combinatorial analysis*. There are two ways of arranging things. The first way is called *permutations* and refers to arranging objects when the order of the objects matters. When the order of the objects does not matter, the arrangements are called *combinations*. We consider permutations first, then combinations in the next subsection.

We want to determine the number of ways n distinct objects can be arranged r ways noting that the order of the objects matters. It may help to think of arranging n distinct pegs in r holes. For the first hole there is a choice of n pegs. With one peg placed in the hole, there is a choice of only $n - 1$ pegs to be placed in the second hole. Since for each of the n choices of pegs for the first hole there are $n - 1$ choices for the second hole, the total number of ways of choosing two pegs to fill two holes is $n(n - 1)$. With two pegs placed in two holes, there are $n - 2$ pegs left to choose for the third hole, so that the number of ways of choosing pegs to fill three holes is $n(n - 1)(n - 2)$, Clearly, the number of ways of choosing n pegs to fill r holes is $n(n - 1)(n - 2)\ldots(n - r + 1)$. We denote the number of ways of arranging n objects r ways by $_nP_r$, hence

$$_nP_r = n(n - 1)(n - 2)\ldots(n - r + 1) \tag{5.7}$$

For $r = n$ we get

$$_nP_n = n(n - 1)(n - 2)\ldots(1) = n!$$

We arrive at the important result that the number of ways of arranging n objects is $n!$. It is worth highlighting this result

Number of ways of arranging n objects	$= n!$

It is convenient to write Eq. (5.7) in terms of factorials. To do this we multiply the numerator and denominator of Eq. (5.7) by $(n - r)!$

$$_nP_r = n(n-1)(n-2)\ldots(n-r+1) = \frac{n(n-1)(n-2)\ldots(n-r+1)(n-r)!}{(n-r)!} = \frac{n!}{(n-r)!}.$$

Let us highlight this important formula:

Number of ways of arranging n objects r ways, or number of **permutations** of n objects r ways	$_nP_r = \dfrac{n!}{(n-r)!}$	(5.8)

Note that for $r = n$, $(n-r)! = (n-n)! = 0! = 1$.

Example 5.2.16 (Permutations)
How many arrangements or permutations of 4 letters can be made from the letters of the word 'permutations'?
Solution
Since there are 12 letters in the word 'permutations' we are permuting 4 letters chosen from 12 letters so that $r = 4$ and $n = 12$. The number of permutations is

$$_nP_r = \frac{n!}{(n-r)} = \frac{12!}{(12-4)!} = \frac{12!}{8!} = \frac{12 \cdot 11 \cdot 10 \cdot 9 \cdot 8!}{8!} = 11,880.$$

5.2.5 Combinations

With permutations the order of the objects matters so, for example, the arrangements of letters *pqrs* and *sprq* are different permutations. For combinations, the order of objects doesn't matter, so that the letter arrangements *pqrs* and *sprq* are considered to be the same combination.

Since r objects can be arranged $r!$ ways, the number of combinations of n objects r ways can be found by replacing these $r!$ arrangements by a single arrangement in the set of permutations of n objects r ways. This can be done by dividing the number of permutations $_nP_r$ by $r!$. We thus obtain the formula for the number of combinations $_nC_r$ by dividing the formula for $_nP_r$ by $r!$:

Number of **combinations** of n objects r ways	$_nC_r = \dfrac{_nP_r}{r!} = \dfrac{n!}{r!(n-r)!}$	(5.9)

Example 5.2.17 (Combinations)
Consider again the arrangement of the letters of the word 'permutations' in groups of 4 letters, but this time the order of the same 4 letters does not matter, which means we are dealing with combinations of the 4 letters. Calculate how many combinations of 4 letters can be made from the word 'permutations'.

Solution
Since there are 12 letters combined four ways, $r = 4$ and $n = 12$. The number of combinations is

$$_nC_r = \frac{n!}{r!(n-r)!} = \frac{12!}{4!(12-4)!} = \frac{12!}{4! \cdot 8!} = \frac{12 \cdot 11 \cdot 10 \cdot 9}{4 \cdot 3 \cdot 2 \cdot 1} = 495.$$

We see that there are vastly fewer combinations than permutations.

Another common notation for combinations is $\binom{n}{r}$, i.e.,

$$\binom{n}{r} = {}_nC_r = \frac{n!}{r!(n-r)!}. \tag{5.10}$$

The numbers given in Eq. (5.10) are called *binomial coefficients* since they appear as the coefficients in the *binomial expansion* $(a + b)^n$.

Example 5.2.18 (Use of combinations and permutations in the calculation of probabilities)
Three balls are picked without replacement from a bag containing 7 red balls and 5 blue balls. What is the probability of picking 1 red ball and two blue balls?

Solution
We can solve the this problem by two methods:
Method 1. Use of permutations
In this method, we consider that the order in which the three balls are chosen matters. In this case the sample space is given by $12 \cdot 11 \cdot 10 = 1320$. We can also use the permutation formula to find the total number of sample points:

Number of sample points = Number of ways of selecting 3 balls from 12 balls

$$= \frac{n!}{(n-r)!} = {}_nP_r = {}_{12}P_3 = \frac{12!}{(12-3)!} = \frac{12!}{9!} = 1320$$

Next, we need to determine the number of desired outcomes, namely, picking 1 red ball and 2 blue balls when the order matters. Since order matters we need to consider the number of ways picking the balls in the order *RBB*, where *RBB* = 1st picked ball is red, 2nd picked ball is blue and 3rd picked ball is blue. This number can be found by noting that the number of ways of picking a red ball first is 7. For each of the red balls picked there are 5 ways of picking the second ball blue, making a total of $7 \cdot 5 = 35$ ways. To each of these 35 ways of picking a red ball first and a blue ball second, there are 4 ways of picking a third ball blue, which makes a total of $7 \cdot 5 \cdot 4 = 140$ ways of picking three balls in the order *RBB*. By the same reasoning, we find that the orders *BRB* and *BBR* can also each be chosen 140 ways, making the total number of desired outcomes $140 + 140 + 140 = 420$.

Rather than calculating by hand, we can use the formula for permutations to calculate the number of ways of picking the orders *RBB*, *BRB* and *BBR* when the order matters. We illustrate this method for *RBB*:

Number of ways of picking *RBB*

= (number of ways of picking 1 red ball from 7 red balls)

· (number of ways of picking 2 blue balls from 5 blue balls)

$$= {}_7P_1 \cdot {}_5P_2 = \frac{7!}{(7-1)!} \cdot \frac{5!}{(5-2)!} = 7 \cdot 5 \cdot 4 = 140$$

The probability $P(1R, 2B)$ of picking 1 red ball and 2 blue balls is thus

$$P(1R, 2B) = \frac{\text{number of ways of picking 1 red ball and 2 blue balls}}{\text{number of sample points}}$$

$$= \frac{3 \cdot \frac{7!}{(7-1)!} \cdot \frac{5!}{(5-2)!}}{\frac{12!}{(12-3)!}} = \frac{420}{1320} = \frac{7}{22},$$

where the abbreviated notation is fully written as $P(1R, 2B) = P(R \cap B \cap B) + P(B \cap R \cap B) + P(B \cap B \cap R)$.

Method 2. Use of combinations

We can also solve the problem by disregarding the order, and therefore use the formula for combinations. This time, the number of sample points is much smaller and given by the formula for combinations:

$$_nC_r = \binom{n}{r} = {}_{12}C_3 = \binom{12}{3} = \frac{12!}{(12-3)!3!} = \frac{12!}{9!3!} = 220$$

The number of desired outcomes is found by the following reasoning, keeping in mind that order doesn't matter: The number of ways of picking a red ball first is 7. To each choice of the 7 red balls there are 5 ways of picking a blue ball second, and 4 ways of picking a blue ball third, giving 20 ways of choosing 2 blue balls for each choice of red ball. But, this would be wrong. Because the order of the two blue balls doesn't matter we must divide by 2 so that to each of the 7 choices of red ball there are only 10 ways to pick two blue balls. Thus, the total number of ways of picking 1 red ball and two blue balls is $7 \cdot 10 = 70$. Similarly, since the order doesn't matter, we do not multiply by 3 as we did for permutations.

As for permutations, instead of calculating by hand, we can use the combinations formula to find the number of desired outcomes:

Number of ways of picking 1 red ball and two blue balls in any order

= (number of ways of picking 1 red ball from 7 red balls in any order)

· (number of ways of picking 2 blue balls from 5 blue balls in any order)

$$= {_7}C_1 \cdot {_5}C_2 = \binom{7}{1}\binom{5}{2} = 70$$

The probability $P(1R, 3B)$ of picking 1 red ball and 2 blue balls is now given by

$$P(1R, 3B) = \frac{\binom{7}{1}\binom{5}{2}}{\binom{12}{3}} = \frac{70}{220} = \frac{7}{22},$$

which, of course, agrees with the answer above calculated using permutations.

Note: As indicated by the definition of our shorthand notation, the probability asked for above is not the same as the probability $P(R \cap B \cap B)$. The latter asks for the probability that the balls are picked in a particular order, whereas the problem we solved above asks for the probability that the balls are picked in any order. We can check this by calculating $P(R \cap B \cap B)$ using combinations and by the use of Theorem 5.2.8:

$$P(R \cap B \cap B) = \frac{{_7}P_1 \, {_5}P_2}{{_{12}}P_3} = \frac{140}{1320} = \frac{7}{66}.$$

By Theorem 5.2.8

$$P(R \cap B \cap B) = P(R)P(B|R)P(B|R \cap B) = \frac{7}{12} \cdot \frac{5}{11} \cdot \frac{4}{10} = \frac{140}{1320} = \frac{7}{66}.$$

Contrast this result with the probability calculated above, noting that $P(R \cap B \cap B) = P(B \cap R \cap B) = P(B \cap B \cap R)$:

$$P(1R, 2B) = P(R \cap B \cap B) + P(B \cap R \cap B) + P(B \cap B \cap R) = \frac{7}{66} + \frac{7}{66} + \frac{7}{66} = \frac{7}{22},$$

which is 3 times larger.

5.3 Probability Distributions

We begin by first considering the definition of a *random variable* as it is an essential concept for what follows.

5.3.1   Random Variables

Random or *stochastic variables* are either numbers forming the sample space of a random experiment (e.g. heights of students in a class) or else they are numbers assigned to each sample point according to some rule. As an example of the latter, consider the sample space $\{HH, HT, TH, TT\}$ resulting from tossing a coin twice. The rule *number of heads in each outcome* assigns the numbers $\{2, 1, 1, 0\}$ to the sample space $\{HH, HT, TH, TT\}$. The rule therefore defines a random variable X having values $\{0, 1, 2\}$. Note that random variables are usually denoted by a capital letter. Since the values of X are discrete, we call X a *discrete random variable*. Where random experiments yield continuous numbers or where continuous numbers are associated with sample points, we call X a *continuous random variable*.

5.3.2   Discrete Probability Distributions

Let X be a discrete random variable with values $x_1, x_2, x_3 \ldots$. The probability $P(X = x_i)$ that X has the value x_i can be written as

$$P(X = x_i) = p(x_i) \tag{5.11}$$

or as

$$P(X = x_i) = p(x) \tag{5.12}$$

if we define $p(x)$, called a *discrete probability distribution function* or a *discrete probability density function* or probability density for short, as follows:

Discrete probabilty density	$p(x) = \begin{cases} p(x_i) & \text{for } x = x_i \\ 0 & \text{for } x \neq x_i \end{cases}$

The names are often shortened to *probability density* or *probability function* . For a function to be a probability function it must satisfy the following conditions:

Conditions for a discrete probability density	1. $0 \leq p(x) \leq 1$ 2. $\sum_x p(x) = 1$	(5.13)

A discrete probability density can be represented by a *bar chart* or *histogram*. Fig. 5.1 shows a bar chart of a probability density and a line graph of a cumulative distribution function (see next section).

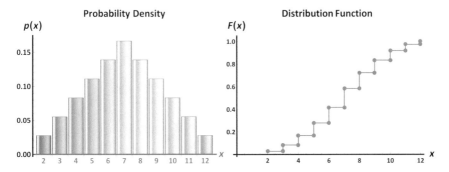

Fig. 5.1 Bar chart of the probability density $p(x)$ and a line graph of the distribution function $F(x)$ of Example 5.3.21

Example 5.3.19 (Probability density)

For the sample space $\{HH, HT, TH, TT\}$ produced by tossing a coin twice, we saw that the variable $X = number\ of\ heads$ takes on the values $X = 0, 1, 2$. Write down the probability density for X.

Solution

It is easy to see that the probabilities associated with these values are $P(X = 0) = \frac{1}{4}$, $P(X = 1) = \frac{1}{2}$ and $P(X = 2) = \frac{1}{4}$. The probability density is defined by these values, i.e., $p(0) = \frac{1}{4}$, $p(1) = \frac{1}{2}$ and $p(2) = \frac{1}{4}$.

5.3.3 *Distribution Functions for Discrete Random Variables*

Instead of asking for the probability of a particular value x_i, we can ask for the probability that X has a value less than some x, where x is continuous with values from $-\infty$ to $+\infty$. This probability is given by the *cumulative distribution function*, usually shortened to *distribution function*, defined by

Cumulative distribution function	$F(x) = P(X \leq x)$

For a discrete random variable the probability is easily obtained by simply adding the probabilities for all values of x_i less than x, i.e.,

$$F(x) = \sum_i p(x_i), \quad x_i < x.$$

Or, it can be obtained from the probability density:

$$F(x) = \sum_{y \leq x} p(y).$$

When X can take only values x_1 to x_n, the distribution function is defined as follows:

$$F(x) = \begin{cases} 0 & \text{for } -\infty < x < x_1 \\ p(x_1) & \text{for } x_1 \leq x < x_2 \\ p(x_1) + p(x_2) & \text{for } x_2 \leq x < x_3 \\ \vdots & \\ p(x_1) + \ldots + p(x_n) & \text{for } x_n \leq x < \infty. \end{cases}$$

We note that since the values of the distribution function $F(x)$ with increasing x are obtained by adding positive or zero probabilities it either increases or remains the same. The distribution function is therefore a *monotonically increasing function*.

Example 5.3.20 (Distribution function)
Write down the distribution function corresponding to the probability function of Example 5.3.19.
Solution
The distribution function corresponding to the probability function of Example 5.3.19 is

$$F(x) = \begin{cases} 0 & \text{for } -\infty < x < 0 \\ \frac{1}{4} & \text{for } 0 \leq x < 1 \\ \frac{1}{4} + \frac{1}{2} = \frac{3}{4} & \text{for } 1 \leq x < 2 \\ \frac{1}{4} + \frac{1}{2} + \frac{1}{4} = 1 & \text{for } 2 \leq x < \infty \end{cases}$$

We note that the jumps $\frac{1}{4} - 0 = \frac{1}{4}$, $\frac{3}{4} - \frac{1}{4} = \frac{1}{2}$ and $1 - \frac{3}{4} = \frac{1}{4}$ in the values of the distribution function $F(x)$ are the probabilities of getting a 'head' in two tosses of a coin. This feature allows the probability density $p(x)$ to be obtained from the distribution function $F(x)$.

Example 5.3.21 (Probability density and distribution function)
Let the discrete random variable X = *sum of the two numbers when two dice are thrown*. Determine the probability density $p(x)$ and distribution function $F(x)$ for X.

Solution

For each of the 6 numbers on the first dice, there corresponds 6 numbers on the second dice so that the total number of pairs of numbers, the sample space, is $6 \cdot 6 = 36$. The probability density is given by the probability that the sum of two numbers is equal to 2, 3,...,12. The number of ways of getting 2 is 1 since the only combination is 1+1=2, the number of ways of getting 3 is 2 since the allowed combinations are 1+2=3 or 2+1=3, and the number of ways of getting 4 is 3 since the allowed combinations are 1+3, 3+1, 2+2, and so on. Thus, the probabilities are $P(X = 2) = \frac{1}{36}$, $P(X = 3) = \frac{2}{36}$, $P(X = 4) = \frac{3}{36}$ etc.. The probability $F(X)$ of getting $X = 2$ or less is $F(2) = \frac{1}{36}$, of getting $X = 3$ or less it's $F(2) = \frac{3}{36}$, of getting $X = 4$ or less it's $F(2) = \frac{7}{36}$ and so on. The table below shows the complete definitions of $p(x)$ and $F(x)$:

x	2	3	4	5	6	7	8	9	10	11	12
$p(x)$	1/36	2/36	3/36	4/36	5/36	6/36	5/36	4/36	3/36	2/36	1/36
$F(x)$	1/36	3/36	6/36	10/36	15/36	21/36	26/36	30/36	33/36	35/36	36/36

A bar chart of $p(x)$ is shown in Fig. 5.1. It has a typical bell shape or Gaussian distribution (see subsection 5.6.4), which describes the distributions of many commonly occurring random variables. A line graph of $F(x)$ is also shown in Fig. 5.1. The graph of $F(x)$ has the staircase function or step function shape typical of discrete distribution functions. Notice, that the jumps correspond to the probabilities $p(x)$. This is easily seen by noting that the jumps are the differences between adjacent values of $F(x)$ and that these differences are the probabilities $p(x)$. For example

$$P(X = 6) = F(X \leq 6) - F(X \leq 5) = \frac{15}{36} - \frac{10}{36} = \frac{5}{36}.$$

5.3.4 *Continuous Probability Density Functions*

If we ask for the probability that a continuous random variable X with values ranging from $x = -\infty$ to $x = +\infty$ has a specific value $X = x$, we will get $P(X = x) = \frac{x}{\infty} = 0$. This is clearly not meaningful, so instead, for a continuous random variable we ask for the probability that it lies in some interval $a \leq x \leq b$. This procedure allows the definition of a continuous probability density $p(x)$. A continuous probability density is defined by the requirement that it satisfies the following conditions, analogues of the discrete case:

Conditions for a **continuous** **probability density function**	1. $0 \leq p(x) \leq 1$ 2. $\int_{-\infty}^{\infty} p(x)\, dx = 1$	(5.14)

The first condition is necessary because a negative probability has no obvious meaning, while the second condition reflects the certainty that the value of a continuous random variable must lie between $-\infty$ to $+\infty$.

Asking for the probability that a banana picked at random from a banana plantation has length between 14 cm to 16 cm is an example of a trial or random experiment which yields a continuous random variable; the length of a banana. Another example, is selecting a student from a class and asking for the probability that her height lies between 1.2 m and 1.5 m.

The probability that a random variable X lies in the interval $a \leq x \leq b$ is given by

Probability that X lies in the interval $a \leq x \leq b$	$P(a \leq x \leq b) = \int_a^b p(x)\, dx$	(5.15)

5.3.5 *Continuous Distribution Function*

We can ask for the probability that a continuous random variable X has a value less than x. This probability is given by a *continuous distribution function $F(x)$*, the analogue of the discrete case, defined by

Continuous distribution function	$F(x) = P(X \leq x) = P(-\infty \leq X \leq x) = \int_{-\infty}^{x} p(v)\, dv$	(5.16)

Because the probability of a specific value x is zero, and as long as $p(x)$ is continuous, '$\leq$' in the above definition is interchangeable with '$<$'. The probability that X lies in the interval $a \leq X \leq b$ is given by the difference $F(b) - F(a)$, i.e.,

$$P(a \leq X \leq b) = F(b) - F(a) = \int_a^b p(x)\, dx. \qquad (5.17)$$

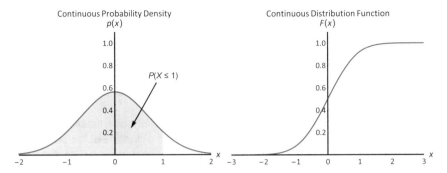

Fig. 5.2 Plot of the continuous probability density $p(x) = \frac{1}{\sqrt{\pi}}e^{-x^2}$ and its distribution function $F(x) = \frac{1}{2}[(1 + \mathrm{erf}(x))]$. The shaded area in the probability function graph shows $P(X \leq 1)$

In Fig. 5.2 the shaded area of the probability density graph gives the probability that $X \leq 1$. The second graph shows the corresponding distribution function.

Example 5.3.22 (Continuous Distribution Function)
The random variable X is described by the probability density

$$p(x) = \begin{cases} 0 \text{ for } x < -1 \\ 0.3(2 - x^2) \text{ for } -1 \leq x \leq 1 \\ 0 \text{ for } x > 1. \end{cases}$$

Find (a) the distribution function, and (b) the probability $P(-0.5 \leq X \leq 0.5)$.
Solution
(a) To find $F(x)$, apply Eq. (5.16) to each of the three intervals in which $p(x)$ is defined:

$$F(x) = \begin{cases} \int_{-\infty}^{-1} 0 \, dx = 0 \text{ for } x < -1 \\ 0.3 \int_{-1}^{x} (2 - v^2) \, dv = 0.6x - 0.1x^3 + 0.5 \text{ for } -1 \leq x \leq 1 \\ \int_{1}^{\infty} 0 \, dx = 0 \text{ for } x > 1. \end{cases}$$

(b) The probability $P(-0.5 \leq X \leq 0.5)$ is found from $F(x)$ found in part (a)

$$P(-0.5 \leq X \leq 0.5) = F(0.5) - F(-0.5) = 0.575$$

It can also be found from Eq. (5.15)

$$P(-0.5 \leq 0.5) = 0.3 \int_{-0.5}^{0.5} (2 - x^2)\, dx = 0.3 \left[2x - \frac{x^3}{3} \right]_{-0.5}^{0.5} = 0.575$$

By considering the probability that X lies in the infinitesimal interval $x \leq X \leq x + dx$ we are led to an important relation, namely, the derivative of the probability density function F is equal to the probability density. The proof is straightforward:

$$F(x) = \int_{x}^{x+dx} p(y)\, dy = p(x)dx \tag{5.18}$$

The last step follows by the definition of an integral as the area under the curve. Here, the area under the curve with projection on the x-axis of length dx is $p(x)dx$. The required result follows immediately from Eq. (5.18):

$$\boxed{\frac{dF(x)}{dx} = p(x)}$$

5.3.6 Joint Distributions

We come now to consider two random variables, either both discrete or both continuous. Generalisation to more than two variables or to mixtures (some random variables discrete, some continuous) is straightforward.

Discrete Joint Distributions

Let X and Y be two discrete random variables which can take discrete values $x_1, x_2, x_3 \ldots$ and $y_1, y_2, y_3 \ldots$ respectively. The probability $P(X = x_i, Y = y_j)$ that $X = x_i$ and $Y = y_j$ can be written as

$$P(X = x_i, Y = y_j) = p(x_i, y_j), \tag{5.19}$$

or as

$$P(X = x_i, Y = y_j) = p(x, y), \tag{5.20}$$

if we define $p(x, y)$, called a *joint discrete probability density function* or simply a *joint discrete probability density*, as follows:

Table 5.1 Joint probability table

$X \backslash Y$	y_1	y_2	$\ldots$	y_m	$p_x(x_i)$
x_1	$p(x_1, y_1)$	$p(x_1, y_2)$	$\ldots$	$p(x_1, y_m)$	$p_x(x_1)$
x_2	$p(x_2, y_1)$	$p(x_2, y_2)$	$\ldots$	$p(x_2, y_m)$	$p_x(x_2)$
$\vdots$	$\vdots$	$\vdots$	$\vdots$	$\ldots$	$\ldots$
x_n	$p(x_n, y_1)$	$p(x_n, y_2)$	$\ldots$	$p(x_n, y_m)$	$p_x(x_n)$
$p_y(y_j)$	$p_y(y_1)$	$p_y(y_2)$	$\ldots$	$p_y(y_m)$	1

$$
\boxed{\textbf{Joint discrete}\ \textbf{probability density}} \quad \boxed{p(x, y) = \begin{cases} p(x_i, y_j) & \text{for } x = x_i, y = y_j \\ 0 & \text{for } x \neq x_i, y \neq y_j \end{cases}} \quad (5.21)
$$

For a function to be a probability function it must satisfy the following conditions:

$$
\boxed{\text{Conditions for a } \textbf{Joint discrete}\ \textbf{probability density}} \quad \boxed{\begin{array}{l} 1.\ 0 \leq p(x, y) \leq 1 \\[2mm] 2.\ \sum_x \sum_y p(x, y) = 1 \end{array}}
$$

The probability that $X = x_i$ irrespective of the value of Y is denoted by $P(X = x_i) = p_x(x_i)$. Supposing X and Y have values $X = x_1, x_2, \ldots x_n$ and $Y = y_1, y_2, \ldots, y_m$, the probability $P(X = x_i)$ is given by

$$
P(X = x_i) = p_x(x_i) = \sum_{j=1}^{m} p(x_i, y_j). \tag{5.22}
$$

Similarly, the probability that $P(Y = y_i)$ is given by

$$
P(Y = y_j) = p_y(y_j) = \sum_{i=1}^{n} p(x_i, y_j). \tag{5.23}
$$

The functions $P(X = x_i) = p_x(x_i)$ and $P(Y = y_j) = p_y(y_j)$ are called *marginal probability density functions* (or *marginal probability densities* for short) and, together with the probability density $p(x_i, y_j)$, can be represented by a *joint probability table* (Table 5.1):

We see from the Table 5.1 that summing columns and rows gives the marginal probability densities in the margins of the table, hence the name 'marginal'. Take, for example, the column labeled y_2. We notice that each entry for p contains y_2. Adding these entries gives the marginal probability function $p_y(y_2)$, i.e.,

$$p(x_1, y_2) + p(x_2, y_2) + \ldots + p(x_n, y_2) = p_y(y_2).$$

Notice also that the bottom right hand corner of Table 5.1 gives the sum of all the probabilities:

$$\sum_{i=1}^{n} \sum_{j=1}^{m} p(x_i, y_j) = 1. \tag{5.24}$$

It follows from Eq. (5.24) that

$$\sum_{i=1}^{n} p_x(x_i) = 1 \quad \text{and} \quad \sum_{i=j}^{m} p_y(y_j) = 1,$$

The *joint distribution function* $F(x, y)$ is defined by

$$F(x, y) = P(X \leq x, Y \leq y) = \sum_{u \leq x} \sum_{v \leq y} p(u, v).$$

Example 5.3.23 (Discrete Joint Probabilities)

3 balls are picked from a bag containing 2 red balls, 3 yellow balls and 4 green balls. Let X = *number of red balls chosen*, and Y = *number of yellow balls chosen*. (a) Define the joint probability function $P(X = x, Y = y) = p(x, y)$. (b) Determine the marginal probability density functions $P(X = x) = p_x(x)$ and $P(Y = y) = p_y(y)$. (c) Draw a table to determine the various marginal probabilities and check the answer by finding the total probability.

Solution

(a) The probability density function is defined by calculating the probabilities of the random variables taking on their allowed values, namely, $X = \{0, 1, 2\}$ and $Y = \{0, 1, 2, 3\}$. First, calculate $p(0, 0)$:

$$\begin{aligned}
p(0, 0) &= \frac{\text{number of ways } X = 0, \ Y = 0}{\text{number of sample points}} \\
&= \frac{\text{number of ways of picking 3 green balls from 4 green balls}}{\text{number of sample points}} \\
&= \binom{4}{3} \div \binom{9}{3} = \frac{4}{84}.
\end{aligned}$$

Next, calculate $p(0, 1)$:

$$p(0, 1) = \frac{\text{number of ways } X = 0, \ Y = 1}{\text{number of sample points}}$$

$$= \frac{(\text{no. of ways to pick 2 green balls from 4})(\text{no. of ways to pick 1 yellow ball from 3})}{\text{number of sample points}}.$$

$$= \binom{4}{2}\binom{3}{1} \div \binom{9}{3} = \frac{18}{84}.$$

The remaining probabilities are calculated in a similar way:

$$p(0, 2) = \binom{4}{1}\binom{3}{2} \div \binom{9}{3} = \frac{12}{84}, \quad p(0, 3) = \binom{3}{3} \div \binom{9}{3} = \frac{1}{84}$$

$$p(1, 0) = \binom{4}{2}\binom{2}{1} \div \binom{9}{3} = \frac{12}{84}, \quad p(1, 1) = \binom{2}{1}\binom{3}{1}\binom{4}{1} \div \binom{9}{3} = \frac{24}{84}$$

$$p(1, 2) = \binom{2}{1}\binom{3}{2} \div \binom{9}{3} = \frac{6}{84}, \quad p(2, 0) = \binom{2}{2}\binom{4}{1} \div \binom{9}{3} = \frac{4}{84}$$

$$p(2, 1) = \binom{2}{2}\binom{3}{1} \div \binom{9}{3} = \frac{3}{84}.$$

(b) The marginal probability distribution $p_x(x)$ is given by Eq. (5.22)

$$p_x(0) = p(0, 0) + p(0, 1) + p(0, 2) + p(0, 3) = \frac{4}{84} + \frac{18}{84} + \frac{12}{84} + \frac{1}{84} = \frac{35}{84}$$

$$p_x(1) = p(1, 0) + p(1, 1) + p(1, 2) = \frac{12}{84} + \frac{24}{84} + \frac{6}{84} = \frac{42}{84}$$

$$p_x(2) = p(2, 0) + p(2, 1) = \frac{4}{84} + \frac{3}{84} = \frac{7}{84}.$$

while the marginal probability distribution $p_y(y)$ is given by Eq. (5.23)

$$p_y(0) = p(0, 0) + p(1, 0) + p(2, 0) = \frac{4}{84} + \frac{12}{84} + \frac{4}{84} = \frac{20}{84}$$

$$p_y(1) = p(0, 1) + p(1, 1) + p(2, 1) = \frac{18}{84} + \frac{24}{84} + \frac{3}{84} = \frac{45}{84}$$

$$p_y(2) = p(0, 2) + p(1, 2) = \frac{12}{84} + \frac{6}{84} = \frac{18}{84}$$

$$p_y(0) = p(0, 3) = \frac{1}{84}.$$

(c) The table of the marginal probabilities and the total probability is given in Table 5.2.

Table 5.2 Joint probability table for Example 5.3.23

$X \backslash Y$	$y_1 = 0$	$y_2 = 1$	$y_3 = 2$	$y_4 = 3$	$p_x(x_i)$
$x_1 = 0$	$p(0,0) = \frac{4}{84}$	$p(0,1) = \frac{18}{84}$	$p(0,2) = \frac{12}{84}$	$p(0,3) = \frac{1}{84}$	$p_x(0) = \frac{35}{84}$
$x_2 = 1$	$p(1,0) = \frac{12}{84}$	$p(1,1) = \frac{24}{84}$	$p(1,2) = \frac{6}{84}$	–	$p_x(1) = \frac{42}{84}$
$x_3 = 2$	$p(2,0) = \frac{4}{84}$	$p(2,1) = \frac{3}{84}$	–	–	$p_x(2) = \frac{7}{84}$
$p_y(y_j)$	$p_y(0) = \frac{20}{84}$	$p_y(1) = \frac{45}{84}$	$p_y(2) = \frac{18}{84}$	$p_y(3) = \frac{1}{84}$	1

Continuous Joint Probability Density

Similarly to the continuous single random variable case, $p(x, y)$ is also called a *joint probability density*. The generalisation from the discrete case is performed in a straightforward way by replacing sums by integrals. To be a probability density, $p(x, y)$ must satisfy conditions analogous to those for the discrete case:

Conditions for a **joint continuous probability density**	1. $0 \leq p(x, y) \leq 1$ 2. $\int_{-\infty}^{\infty} \int_{-\infty}^{\infty} p(x, y)\, dxdy = 1$

$$(5.25)$$

With these conditions satisfied, the probability $P(a < X < b, c < Y < d)$ that X has a value in the interval $a < X < b$ and Y has a value in the interval $c < Y < d$ is

$$P(a < X < b, c < Y < d) = \int_{x=a}^{x=b} \int_{y=c}^{y=d} p(x, y)\, dxdy$$

The *joint distribution function* generalises to

$$F(x, y) = P(X \leq x, Y \leq y) = \int_{u=-\infty}^{u=x} \int_{v=-\infty}^{v=y} p(u, v)\, dudv$$

Following similar steps as in subsection 5.3.5, we can show that the joint probability density is the second partial derivative of the joint distribution function:

$$p(x, y) = \frac{\partial^2 F}{\partial x \partial y} \tag{5.26}$$

Proof of Eq. (5.26): Let X lie in the infinitesimal interval $x \leq X \leq x + \Delta x$, and y in the infinitesimal interval $y \leq Y \leq y + \Delta y$, then

$$P(x \leq X \leq x + \Delta x, y \leq Y \leq y + \Delta y) = F(x, y) = \int_{u=x}^{u=x+\Delta x} \int_{v=y}^{v=y+\Delta y} p(u, v)\, dudv$$

The integral gives the volume under the surface $p(x, y)$ with area projection on the xy-plane $dxdy$ at point (x, y). Since the area $dxdy$ is infinitesimal, this volume is given by $p(x, y)dxdy$, hence

$$dF = p(x, y)dxdy.$$

Rearrangement gives

$$\frac{\partial^2 F}{\partial x \partial y} = p(x, y),$$

which completes the proof.

The *marginal probability density functions* are given by

$$p_x(x) = \int_{y=-\infty}^{\infty} p(x, y)\, dy \quad \text{and} \quad p_y(y) = \int_{x=-\infty}^{\infty} p(x, y)\, dx, \qquad (5.27)$$

while the marginal distribution functions are given by

$$P(X \le x) = F_x(x) = \int_{u=-\infty}^{u=x} \int_{v=-\infty}^{v=\infty} p(u, v)\, du\, dv \qquad (5.28)$$

$$P(Y \le y) = F_y(y) = \int_{u=-\infty}^{u=\infty} \int_{v=-\infty}^{v=y} p(u, v)\, du\, dv. \qquad (5.29)$$

Independent Random Variables

The random variables X and Y are said to be independent when the probability function $p(x, y)$, whether discrete or continuous, factorises into the product $p_x(x)p_y(y)$, i.e.,

Condition for random variables X, Y to be independent	$p(x, y) = p_x(x)p_y(y)$

The factorisation of the distribution functions $F(x, y)$ into the product

$$F(x) = F_x(x)F_y(y)$$

of marginal distribution functions also expresses the statistical independence of X and Y.

When $p(x, y)$ or $F(x, y)$ do not factorise, the random variables X and Y are statistically dependent.

Example 5.3.24 (Continuous Joint Probability Density)
The *continuous joint probability density* for two random variables X and Y is defined by

$$p(x, y) = \begin{cases} e^{-2x}e^{-\frac{y}{2}} & \text{for } 0 \le x \le \infty,\ 0 \le y \le \infty \\ 0 & \text{otherwise.} \end{cases}$$

Determine (a) $P(X \geq 1, Y \leq 1)$, (b) $P(X \leq Y)$ and (c) $P(X \leq a)$.
Solution
(a)

$$P(X \geq 1, Y \leq 1) = \int_{-\infty}^{1} \int_{1}^{\infty} e^{-2x} e^{-\frac{y}{2}} \, dx dy$$

$$= \int_{0}^{1} \int_{1}^{\infty} e^{-2x} e^{-\frac{y}{2}} \, dx dy$$

$$= \int_{0}^{1} e^{-\frac{y}{2}} \left[\frac{1}{-2} e^{-2x} \right]_{1}^{\infty} dy$$

$$= \int_{0}^{1} e^{-\frac{y}{2}} \left[0 - \frac{1}{-2} e^{-2} \right] dy$$

$$= \frac{1}{2} e^{-2} \left[\frac{1}{-\frac{1}{2}} e^{-\frac{y}{2}} \right]_{0}^{1}$$

$$= \frac{1}{2} e^{-2} \left(-2e^{-\frac{1}{2}} + 2 \right)$$

$$= e^{-2} - e^{-\frac{5}{2}}$$

(b)

$$P(X \leq Y) = \int_{0}^{\infty} \int_{0}^{y} e^{-2x} e^{-\frac{y}{2}} \, dx dy$$

$$= \int_{0}^{\infty} e^{-\frac{y}{2}} \left[\frac{1}{-2} e^{-2x} \right]_{0}^{y} dy$$

$$= \int_{0}^{\infty} e^{-\frac{y}{2}} \left[\frac{1}{-2} e^{-2y} - \frac{1}{-2} \right] dy$$

$$= \int_{0}^{\infty} \left(\frac{1}{-2} e^{-\frac{5}{2}y} + \frac{1}{2} e^{-\frac{y}{2}} \right) dy$$

$$= \left[\frac{1}{5} e^{-\frac{5}{2}y} + \frac{1}{-1} e^{-\frac{y}{2}} \right]_{0}^{\infty}$$

$$= -\frac{1}{5} + 1$$

$$= \frac{4}{5}$$

(c)

$$P(X \le a) = \int_0^\infty \int_0^a e^{-2x} e^{-\frac{y}{2}} \, dx dy$$

$$= \int_0^a e^{-2x} \left[\frac{1}{-\frac{1}{2}} e^{-\frac{y}{2}} \right]_0^\infty dx$$

$$= \int_0^a e^{-2x} \left(0 - \frac{1}{-\frac{1}{2}} \right) dx$$

$$= 2 \left[\frac{1}{-2} e^{-2x} \right]_0^a$$

$$= 2 \left(\frac{1}{-2} e^{-2a} - \frac{1}{-2} \right)$$

$$= 1 - e^{2a}$$

Conditional Probability Densities

Formulae for the *conditional probability densities* $P(Y = y|X = x) = p(y|x)$ and $P(X = x|Y = y) = p(x|y)$ are obtained by substituting the probability density $p(x, y)$ and the marginal probability densities $p_x(x, y)$ and $p_y(x, y)$ for the probabilities in the conditional probability formula (5.1). For the discrete case we substitute Eqs. (5.21), (5.22) and (5.23). For the continuous case we substitute Eqs. (5.25) and (5.27). With these equations, the definitions of $P(Y = y|X = x) = p(y|x)$ and $P(X = x|Y = y) = p(x|y)$ take the same form for both the continuous and discrete cases:

$$P(Y = y \mid X = x) = p(y|x) = \frac{p(x, y)}{p_x(x)}. \tag{5.30}$$

Similarly, the probability that $X = x$ given $Y = y$ is

$$P(X = x \mid Y = y) = p(x|y) = \frac{p(x, y)}{p_y(y)}. \tag{5.31}$$

The functions $p(y|x)$ and $p(x|y)$ are called *conditional probability density functions*.

Using Eq. (5.31) with $p(x, y)$ a continuous probability function, the probability $P(a \le X \le b \mid y \le Y \le y + dy)$ that X lies in the interval $a \le X \le b$ given that y lies in the interval $y \le Y \le y + dy$ is given by

$$P(a \le X \le b \mid y \le Y \le y + dy) = \int_a^b p(x|y) \, dx.$$

Since dy is infinitesimal, we can write the formula more simply and interpret it as the probability that $a < X < b$ given $Y = y$:

$$P(a \leq X \leq b \mid Y = y) = \int_a^b p(x|y)\, dx.$$

Note that this is a 'working' formula to find $a \leq X \leq b$ given $Y = y$ since the probability of $Y = 0$ is 0.

Example 5.3.25 (Discrete Conditional Probability Density)
Consider the discrete joint probability density

$$p(0, 0) = 0.3, \quad p(0, 1) = 0.2, \quad p(1, 0) = 0.4, \quad p(1, 1) = 0.1,$$

for the random variables X and Y. Calculate the conditional probability density $p(x|y)$ given that $Y = y = 1$.
Solution
To find $p(x|1)$ we use formula (5.31):

$$p(x|y) = \frac{p(x, y)}{p_y(y)}, \quad p(x|1) = \frac{p(x, 1)}{p_y(1)}$$

First we find $p_y(1)$

$$p_y(1) = p(0, 1) + p(1, 1) = 0.2 + 0.1 = 0.3$$

With this, the conditional probability density $p(x|1)$ is defined by

$$p(0|1) = \frac{p(0, 1)}{p_y(1)} = \frac{0.2}{0.3} = \frac{2}{3}, \quad p(1|1) = \frac{P(1, 1)}{P_Y(1)} = \frac{0.1}{0.3} = \frac{1}{3}$$

Example 5.3.26 (Continuous Conditional Probability Density)
Consider the continuous joint probability density

$$p(x, y) = \begin{cases} \frac{4}{15}(x - 2)(y - 3) & \text{for } 0 \leq x \leq 1,\ 0 \leq y \leq 1 \\ 0 & \text{otherwise.} \end{cases} \tag{5.32}$$

Determine the conditional probability density $p(x|y)$ of getting X given that $Y = y$.

Solution
To find $p(x, y)$ we use formula (5.31), i.e.,

$$p(x|y) = \frac{p(x, y)}{p_y(y)}. \tag{5.33}$$

First find the marginal probability density $p_y(y)$ using formula (5.27)

$$
\begin{aligned}
p_y(y) &= \int_{x=-\infty}^{\infty} p(x, y)\, dx = \frac{4}{15} \int_{x=0}^{1} (x-2)(y-3)\, dx \\
&= \frac{4}{15} \int_{x=0}^{1} (xy - 3x - 2y + 6)\, dx \\
&= \frac{4}{15} \left[\frac{x^2 y}{2} - \frac{3x^2}{2} - 2yx + 6x \right]_0^1 \\
&= \frac{4}{15} \left(\frac{y}{2} - \frac{3}{2} - 2y + 6 \right) = \frac{2}{15}(9 - 3y) \tag{5.34}
\end{aligned}
$$

Substituting Eqs. (5.32) and (5.34) into Eq. (5.33), we get the required conditional probability density $p(x|y)$,

$$
p(x|y) = \begin{cases} \frac{2(x-2)(y-3)}{(9-3y)} & \text{for } 0 \le x \le 1,\ 0 \le y \le 1 \\ 0 & \text{otherwise.} \end{cases} \tag{5.35}
$$

5.4 The Mean and Related Concepts

We saw in earlier chapters the importance of the mean in the analysis of experimental data. It is the best estimate of the true value of a measured quantity (a value that can never be known). More generally, the mean plays a fundamental role in all sorts of statistical applications such as determining the life expectancy of a population for annuity calculations, or determining average weights or heights of a population, etc.. Knowing the mean alone is not enough; the scatter or spread of data is also important. For example, an experimental measurement of a quantity producing a narrow spread of measured values is regarded as more precise (though not necessarily accurate - since the average of the data may not be close to the true value because of the presence of systematic errors) than one that produces a wide spread. The concept of variance, or its square root, the standard deviation, are important measures of spread or scatter of data.

5.4.1 The Mean

The *mean*, also called the average or expectation value, is commonly denoted either with angle brackets $\langle X \rangle$ or by an over-line $\overline{X}$. Another common notation, especially when the mean appears in a formula, is to represent the mean by the Greek letter μ. In this chapter we shall mostly use angle brackets or μ to represent the mean.

For a discrete data set of n quantities $X = x_1, x_2, \ldots, x_n$, the mean is defined as the sum of the quantities divided by the number n of quantities, that is

$$\langle x \rangle = \frac{x_1 + x_2 + x_3 \ldots + x_n}{n} = \frac{\sum_{i=1}^{n} x_i}{n} \tag{5.36}$$

A quantity may occur more than once. Let $f(x_i)$, called the *frequency*, be the number of times a quantity x_i occurs. For example, in

$$\langle x \rangle = \frac{x_1 + 4x_2 + 2x_3 \ldots + x_n}{n}$$

x_2 occurs 4 times so its frequency $f(x_2) = 4$, while x_3 occurs 2 times so its frequency $f(x_3) = 2$. Using this notation, the formula for the mean is better written as

$$\langle X \rangle = \frac{f(x_1)x_1 + f(x_2)x_2 + f(x_3)x_3 \ldots + f(x_n)x_n}{n} = \frac{\sum_{i=1}^{n} f(x_i)x_i}{n} \tag{5.37}$$

Noting that the frequency $f(x_i)$ divided by n is just the probability $p(x_i) = \frac{f(x_i)}{n}$ of occurrence of the quantity x_i, we can also write the formula for the mean as

$$\langle X \rangle = p(x_1)x_1 + p(x_2)x_2 + p(x_3)x_3 \ldots + p(x_n)x_n = \sum_{i=1}^{n} p(x_i)x_i \tag{5.38}$$

Collecting results, we have

Mean of a discrete quantity	$\langle X \rangle = \dfrac{\sum_{i=1}^{n} f(x_i)x_i}{n} = \sum_{i=1}^{n} p(x_i)x_i$ (5.39)

Example 5.4.27 (Mean of a set of discrete quantities I)
A coin is tossed 4 times. The random variable X = *the number of heads after four tosses*. Hence, $X = 0, 1, 2, 3, 4$ with probabilities given by $P(X = 0) = \frac{1}{16}$, $P(X = 1) = \frac{4}{16}$, $P(X = 2) = \frac{6}{16}$, $P(X = 3) = \frac{4}{16}$, $P(X = 4) = \frac{1}{16}$, where the number of sample points is 16. Find the mean $\langle X \rangle$.

Solution
The mean $\langle X \rangle$ is given by

$$\langle X \rangle = \sum_{i=1}^{n} p(x_i)x_i = \sum_{i=1}^{5} P(X = x_i)x_i$$
$$= 0P(X = 0) + 1P(X = 1) + 2P(X = 2) + 3P(X = 3) + 4P(X = 4)$$
$$= 0 \cdot \frac{1}{16} + 1 \cdot \frac{4}{16} + 2 \cdot \frac{6}{16} + 3 \cdot \frac{4}{16} + 4 \cdot \frac{1}{16} = 2$$

Example 5.4.28 (Mean of a set of discrete quantities II)
In an experiment, 10 measurements of the diameter of a copper wire are made, with the results given in millimeters: 0.70, 0.69, 0.68, 0.72, 0.71, 0.69, 0.71, 0.67, 0.73, 0.69. Find the mean.
Solution
The diameter of a wire is a continuous random variable. However, we can only take measurements up to a finite number of decimal places depending on the precision of the instrument (here, a Vernier scale giving measurements in millimeters to 2 decimal places). This means that we can treat length measurements as discrete. Note that since the diameter is a continuous variable, we can also find the mean using the formula for the mean of a continuous random variable after determining a suitable probability density which best fits the distribution of the data. The latter approach is most useful when dealing with large amounts of data. For small amounts of data, as here, it is better to use the discrete formula.

Thus, using the discrete formula (5.39), the mean $\langle d \rangle$ is given by

$$\langle d \rangle = \frac{1 \cdot 0.70 + 3 \cdot 0.69 + 1 \cdot 0.68 + 1 \cdot 0.72 + 2 \cdot 0.71 + 1 \cdot 0.67 + 1 \cdot 0.73}{10}$$
$$= 0.699 \text{ mm} = 0.70 \text{ mm (2 sf)}$$

The mean of a continuous variable is given in terms of the probability density:

| **Mean** of a continuous quantity | $\langle X \rangle = \int_{-\infty}^{\infty} xp(x)\, dx$ | (5.40) |

To motivate the definition we compare it to the discrete case by breaking up the continuous variable x into discrete parts x_i, thereby approximating the integral as the sum of small areas $p(x_i)\Delta x_i$ summed from $i = -\infty$ to ∞. The area $p(x_i)\Delta x_i$ is the probability of finding $X = x$ in the interval x to $x + dx$ and is therefore analogous to the discrete case probability $p(x_i)$. In this case, the continuous mean can be approximated using the discrete mean formula:

$$\langle X \rangle = \sum_{-\infty}^{\infty} [p(x_i) \Delta x_i] x_i \tag{5.41}$$

In the limit $\Delta x_i \to dx$, $\sum_{-\infty}^{\infty} \to \int_{-\infty}^{\infty}$ and Eq. (5.41) reduces to formula (5.40)

$$\langle X \rangle = \int_{-\infty}^{\infty} x p(x) \, dx,$$

Example 5.4.29 (Mean of a continuous quantity)
Find the expectation value of a random variable X given that it is described by the probability density

$$p(x) = \frac{1}{\sqrt{\pi}} e^{-(x-2)^2},$$

which is a Gaussian distribution (see Sect. 5.6.4).
Solution

$$\langle X \rangle = \frac{1}{\sqrt{\pi}} \int_{-\infty}^{\infty} e^{-(x-2)^2} x \, dx = 2$$

Functions of Random Variables

If X is either a discrete or continuous random variable, then any function $g(X)$ is correspondingly also a discrete or continuous random variable. The definitions of the means are similar to Eqs. (5.39) and (5.40):

Mean of a discrete function of a random variable	$\langle g(X) \rangle = \dfrac{\sum_{i=1}^{n} f(x_i) g(x_i)}{n} = \sum_{i=1}^{n} g(x_i) p(x_i)$
$f(x_i) = $ frequency of x_i,	$p(x_i) = $ probability of x_i

$$\tag{5.42}$$

Mean of a continuous function of a random variable	$\langle g(X) \rangle = \int_{-\infty}^{\infty} g(x)p(x)\, dx$
$p(x) =$ probability density	

(5.43)

The generalisation of the definition of the mean to two or more random variables is straightforward. For two random variables X and Y the generalisations for both the discrete and continuous cases are:

Mean of a discrete function of random variables X, Y	$\langle g(X, Y) \rangle = \sum_{i=1}^{n} \sum_{j=1}^{m} g(x_i, y_j)p(x_i, y_j)$
$p(x_i, y_j) =$ probability of x_i and y_j	

Mean of a continuous function of random variables X, Y	$\langle g(X, Y) \rangle = \int_{-\infty}^{\infty} \int_{-\infty}^{\infty} g(x, y)p(x, y)\, dxdy$
$p(x, y) =$ probability density	

Here are some useful properties of the mean:

1. For any constant k	$\langle kX \rangle = k \langle X \rangle$
2. For any random variables X, Y	$\langle X + Y \rangle = \langle X \rangle + \langle Y \rangle$
3. For independent random variables X, Y	$\langle XY \rangle = \langle X \rangle \langle Y \rangle$

Example 5.4.30 (Mean of a function of a random variable)
The following are 10 measured values of one side of a cubical water tank in units of metres: 0.8935, 0.8935, 0.8935, 0.8745, 0.8745, 0.9155, 0.9025, 0.9025, 0.9125, 0.9125. Let the random variable $X =$ *the measured length of one side of the water tank*, and the function of the random variable $g(X) =$ *the volume of the tank* $= X^3$. Find the means of X and $g(X)$.

Solution

As in Example 5.4.28, X is continuous but the measured values are discrete so we can use the discrete mean formula. Using formula (5.39) the mean of X is

$$\langle X \rangle = \frac{3 \cdot 0.8935 + 2 \cdot 0.8745 + 0.9155 + 2 \cdot 0.9025 + 2 \cdot 0.9125}{10} = 0.8975 \text{ m.}$$

Using formula (5.43) the mean of $g(X)$ is

$$\langle g(X) \rangle = \frac{3 \cdot 0.8935^3 + 2 \cdot 0.8745^3 + 0.9155^3 + 2 \cdot 0.9025^3 + 2 \cdot 0.9125^3}{10} = 0.7235 \text{ m}^3.$$

5.4.2 Variance and Standard Deviation

As we mentioned earlier, the mean alone, though important, is not enough. The mean alone gives no indication of the precision of an experiment. For this we need an idea of the spread or scatter of the measured values. The need to measure the spread or scatter, in addition to the mean, is true for any set of data values. Such a measure is the *variance*, or its square root, the *standard deviation*.

We defined the standard deviation σ in Chap. 3 and we repeat the definition here. We will also adopt the commonly used notation μ to represent the mean. Using this notation, the variance for both discrete and continuous random variables is defined by

$$\sigma^2 = \langle (x - \mu)^2 \rangle$$

In words, the variance is found by taking the difference of the mean with all data values, $(x_i - \mu)$, squaring each difference to obtain $(x - \mu)^2$, and then taking the mean of the squared differences. We may recall that in Chap. 3 we called the differences $r = (x_i - \mu)$ residuals. This general definition can be written more specifically for discrete and continuous random variables as follows:

Variance of a discrete quantity	$\sigma^2 = \dfrac{\sum_{i=1}^{n} f(x_i)(x_i - \mu)^2}{n} = \sum_{i=1}^{n} p(x_i)(x_i - \mu)^2$	(5.44)

Variance of a continuous quantity	$\sigma^2 = \int_{-\infty}^{\infty} p(x)(x - \mu)^2 \, dx$	(5.45)

The standard deviation formulae follow by taking the square roots of formulae (5.44) and (5.45):

Standard Deviation of a discrete quantity	$\sigma = \left[\dfrac{\sum_{i=1}^{n} f(x_i)(x_i - \mu)^2}{n} \right]^{\frac{1}{2}} = \left[\sum_{i=1}^{n} p(x_i)(x_i - \mu)^2 \right]^{\frac{1}{2}}$

$$(5.46)$$

Standard Deviation of a continuous quantity	$\sigma = \left[\int_{-\infty}^{\infty} p(x)(x - \mu)^2 \, dx \right]^{\frac{1}{2}}$	(5.47)

In Sect. 3.1 we distinguished between the *population standard deviation* and the *sample standard deviation*. We mentioned there that formula (3.2) for the sample standard deviation is a better estimate for the real (impossible to know) standard deviation than the population standard deviation, Eq. (3.1), when only a portion (sample) of the population is known. This situation arises in cases where the population is discrete but infinite (as is the case of counting particles emitted by radioactive sources in finite time intervals, when the full population requires counting for an infinite time interval), or when the population is continuous (as in measurements of continuous quantities). Actually, even for a finite population, there are numerous situations where the entire population cannot be accessed. For example, in finding the average height of women in a given country, the population is too large to be fully accessed. The important example, mentioned above, that is most relevant to us here, is experimental measurement (length, speed etc.). Quantities such as length are continuous and require an infinite number of measurements to obtain the true (impossible to know) mean. Obviously, this is impossible and all measurements of continuous quantities are finite both in the number of measurements taken and in the number of decimal places of each measurement. Thus, in all such cases where only a portion of the population is available, the best estimate of the true standard deviation is given by a slight generalistion of Eq. (3.1), namely,

$$\sigma = \sqrt{\frac{\sum_{i=1}^{n} f(x_i)(x_i - \mu)^2}{(n-1)}}, \qquad (5.48)$$

which is the *sample standard deviation* also defined earlier in Eq. (3.2). The situation is identical for the variance, with the sample variance given by[5]

[5]Note that many spreadsheet and mathematical computer packages have inbuilt formulae for the mean, variance and standard deviation. However, it is not always stated whether the formulae are for the population or for the sample variance or standard deviation. Where not stated, the formulae are invariably for the sample variance and sample standard deviation.

$$\sigma^2 = \frac{\sum_{i=1}^{n} f(x_i)(x_i - \mu)^2}{(n-1)}. \qquad (5.49)$$

Since all real measurements (even of continuous variables) are discrete, why use continuous probability distributions $p(x)$? The first answer is that continuous probability distributions are much easier to handle mathematically. Even for discrete populations, when they are very large, it is convenient to approximate them with a continuous probability distribution. A second more important reason is that continuous probability distributions can be viewed as the infinite limits of probability distributions resulting from the finite measurements of continuous quantities. As such, it is argued that statistical quantities such as the mean or the variance calculated from these distributions are the best estimates of these quantities.

An important formula for variance for the discrete case is derived as follows:

$$\sigma^2 = \langle (x_i - \mu)^2 \rangle = \sum_{i=1}^{n} (x_i - \mu)^2 p(x_i) = \sum_{i=1}^{n} (x_i^2 - 2x_i\mu + \mu^2) p(x_i)$$

$$= \sum_{i=1}^{n} x_i^2 p(x_i) - 2\mu \sum_{i=1}^{n} x_i p(x_i) + \mu^2 \sum_{i=1}^{n} p(x_i) = \langle x^2 \rangle - 2\mu^2 + \mu^2$$

$$= \langle x^2 \rangle - \mu^2 = \langle x^2 \rangle - \langle x \rangle^2,$$

where we have used

$$\sum_{i=1}^{n} x_i p(x_i) = \mu \quad \text{and} \quad \sum_{i=1}^{n} p(x_i) = 1.$$

The same formula applies to the continuous case as we now show:

$$\sigma^2 = \langle (x_i - \mu)^2 \rangle = \int_{-\infty}^{\infty} p(x)(x - \mu)^2 \, dx = \int_{-\infty}^{\infty} p(x)(x^2 - 2x\mu + \mu^2) \, dx$$

$$= \int_{-\infty}^{\infty} p(x)x^2 \, dx - 2\mu \int_{-\infty}^{\infty} p(x)x \, dx + \mu^2 \int_{-\infty}^{\infty} p(x) \, dx = \langle x^2 \rangle - 2\mu^2 + \mu^2,$$

$$= \langle x^2 \rangle - \langle x \rangle^2,$$

where, again, we have used the definition of the mean and the requirement that probabilities must add to 1. This gives us an important formula for standard deviation for both the discrete and continuous case, noting that x can be either discrete or continuous:

Fig. 5.3 Plots of a large and small variance

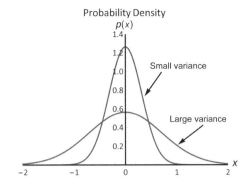

Another formula for variance | $\sigma^2 = \langle x^2 \rangle - \langle x \rangle^2$ | (5.50)

Figure 5.3 shows that a large standard deviation indicates a large spread of data values, while a small standard deviation shows a small spread of data values.

Here are some useful properties of the variance which apply both to discrete and to continuous random variables X and Y:

1.	$\sigma^2 = \langle (x - \mu)^2 \rangle = \langle x^2 \rangle - \mu^2$
2. For any constant k	$\sigma^2_{kX} = k^2 \sigma_X$
3. For independent random variables X, Y	$\sigma^2_{X \pm Y} = \sigma^2_X \pm \sigma^2_Y$

Notation: σ^2_{kX} = variance of the product $kX = \{kx_i\}$ for the discrete case, $kX = kx$ for the continuous case and $\sigma^2_{X \pm Y}$ = variance of $X \pm Y$. These properties can be generalised to more than two random variables in obvious ways.

Example 5.4.31 (Variance and standard deviation of a set of discrete quantities)
Let the random variable X be the outcome of throwing a fair dice. Find the variance and the standard deviation.
Solution
Each number has the same probability $\frac{1}{6}$ of being obtained, i.e.,

$$p(1) = p(2) = p(3) = p(4) = p(5) = p(6) = \frac{1}{6}.$$

First, find the mean

$$\langle X \rangle = \mu = \left(\frac{1}{6}\right)1 + \left(\frac{1}{6}\right)2 + \left(\frac{1}{6}\right)3 + \left(\frac{1}{6}\right)4 + \left(\frac{1}{6}\right)5 + \left(\frac{1}{6}\right)6 = 3.5$$

The variance is given by formula (5.44)

$$\sigma^2 = \frac{1}{6}(1 - 3.5)^2 + \frac{1}{6}(2 - 3.5)^2 + \frac{1}{6}(3 - 3.5)^2 + \frac{1}{6}(4 - 3.5)^2 + \frac{1}{6}(5 - 3.5)^2 + \frac{1}{6}(6 - 3.5)^2$$

$$= \frac{35}{12} = 2.917. \tag{5.51}$$

The standard deviation is

$$\sigma = \sqrt{2.917} = 1.708$$

The variance may also be found from formula (5.50). First find $\langle X^2 \rangle$:

$$\langle X^2 \rangle = \frac{1}{6}(1^2 + 2^2 + 3^2 + 4^2 + 5^2 + 6^2) = \frac{91}{6}.$$

Substituting $\langle X^2 \rangle$ and $\langle X \rangle$ into formula (5.50) we get

$$\sigma^2 = \langle X^2 \rangle - \langle X \rangle^2 = \frac{91}{6} - \left(\frac{7}{2}\right)^2 = \frac{35}{12} = 2.917,$$

as before.

Example 5.4.32 (Variance and standard deviation of a continuous quantity)
Find the variance and standard deviation of a random variable X described by the probability density

$$p(x, y) = \begin{cases} 3x^2 & \text{for } 0 \le x \le 1 \\ 0 & \text{otherwise.} \end{cases}$$

Solution
First find the mean using formula (5.40)

$$\langle X \rangle = \int_0^1 3x^2 x \, dx = 0.75.$$

Use formula (5.45) to find the variance

$$\sigma^2 = \int_0^1 3x^2 (x - 0.75)^2 \, dx = 0.03750.$$

The standard deviation is

$$\sigma = 0.1936.$$

It is sometimes convenient to define a *standardised random variable* such that it has mean $\mu = 0$ and standard deviation $\sigma = 1$. If X is a random variable with mean μ and standard deviation σ, the corresponding standardised random variable Z is given by

Standardised random variable Z of X	$Z = \dfrac{X - \mu}{\sigma}$
Mean of Z	$\mu_Z = 0$
Standard deviation of Z	$\sigma_Z = 1$

Standardised random variables are dimensionless and are useful for comparing different distributions.

We mention, in passing, that the concept of variance can be generalised by taking the rth power of the residues. The quantity so obtained is called the rth moment. For both discrete and continuous random variables the rth moment is defined by

rth moment of X	$\mu_r = \langle (x - \mu)^r \rangle$

Also in passing, we give the definition of the *moment generating function $M(t)$* given by the mean of the function e^{tX} of the random variable X, with t a parameter:

$$M(t) = \langle e^{tX} \rangle = \begin{cases} \sum_i^n e^{tx_i} p(x_i), & X \text{ discrete with probability function } p(x_i) \\ \int_{-\infty}^{\infty} e^{tx} p(x) \, dx, & X \text{ continuous with probability function } p(x). \end{cases}$$

$$(5.52)$$

Mean, Variance and Covariance of Joint Distributions

The definitions of mean and variance of two random variables X and Y with probability density $p(x, y)$ can be generalized in an obvious way. The mean and variance of X are denoted by μ_X and σ_X^2 respectively, while those for Y are denoted by μ_Y and σ_Y^2. First, the various definitions for the discrete case:

Mean of discrete X	$\mu_X = \sum_{i=1}^{n} \sum_{j=1}^{m} x_i p(x_i, y_j)$	
Variance of discrete X	$\sigma_X^2 = \langle (x - \mu_X)^2 \rangle = \sum_{i=1}^{n} \sum_{j=1}^{m} (x_i - \mu_X)^2 p(x_i, y_j)$	(5.53)
Mean of discrete Y	$\mu_Y = \sum_{i=1}^{n} \sum_{j=1}^{m} y_i p(x_i, y_j)$	
Variance of discrete Y	$\sigma_Y^2 = \langle (x - \mu_Y)^2 \rangle = \sum_{i=1}^{n} \sum_{j=1}^{m} (x_i - \mu_Y)^2 p(x_i, y_j)$	

Next we give the various definitions for the continuous case:

Mean of continuous X	$\mu_X = \int_{-\infty}^{\infty} \int_{-\infty}^{\infty} x p(x, y) \, dxdy$
Variance of continuous X	$\sigma_X^2 = \langle (x - \mu_X)^2 \rangle = \int_{-\infty}^{\infty} \int_{-\infty}^{\infty} (x - \mu_X)^2 p(x, y) \, dxdy$
Mean of continuous Y	$\mu_Y = \int_{-\infty}^{\infty} \int_{-\infty}^{\infty} y p(x, y) \, dxdy$
Variance of continuous Y	$\sigma_Y^2 = \langle (x - \mu_Y)^2 \rangle = \int_{-\infty}^{\infty} \int_{-\infty}^{\infty} (x - \mu_Y)^2 p(x, y) \, dxdy$

$$(5.54)$$

With joint distributions, another important quantity arises, namely, the *covariance*. It is denoted by σ_{xy} and has the general definition:

Covariance of X, Y	$\sigma_{XY} = \langle (x - \mu_X)(y - \mu_Y) \rangle$

This definition may also be written more specifically for the discrete and continuous cases as follows:

Covariance of discrete X, Y	$\sigma_{XY} = \sum_{i=1}^{n} \sum_{j=1}^{m} (x_i - \mu_X)(y_i - \mu_Y) p(x_i, y_j)$	(5.55)
Covariance of continuous X, Y	$\sigma_{XY} = \int_{-\infty}^{\infty} \int_{-\infty}^{\infty} (x - \mu_X)(y - \mu_Y) p(x, y) \, dxdy$	(5.56)

The importance of covariance is that it indicates the extent to which the random variables X and Y depend on each other. This dependence is made more precise by the correlation coefficient defined in the next section.

Some properties of the covariance:

1.	$\sigma_{XY} = \mu_{XY} - \mu_X \mu_Y$
2. For independent random variables X, Y	$\sigma_{XY} = 0$
3.	$\sigma_{X\pm Y}^2 = \sigma_X^2 + \sigma_Y^2 \pm 2\sigma_{XY}$
4.	$|\sigma_{XY}| \leq \sigma_X \sigma_Y$

Correlation Coefficient

In joint distributions, the variables may be completely independent, in which case $\sigma_{XY} = 0$, or completely dependent, for example, when $X = Y$, in which case $\sigma_{XY} = \sigma_X \sigma_Y$. But in many cases, X and Y are partially dependent on each other, i.e., they are correlated to some extent. We need a way to measure the degree of correlation. Such a measure is the *correlation coefficient* denoted by ρ and defined by

Definition of the **Correlation Coefficient**	$\rho = \dfrac{\sigma_{XY}}{\sigma_X \, \sigma_Y}$	(5.57)

The correlation coefficient can take on values in the interval $-1 \leq \rho \leq 1$. When $\rho = 0$, X and Y are said to be uncorrelated, otherwise they are correlated.

Example 5.4.33 (Variance and standard deviation of a discrete joint distribution) Consider again example 5.3.23 in which 3 balls are picked from a bag containing 2 red balls, 3 yellow balls and 4 green balls. We found the following joint probabilities:

$$p(0, 0) = \frac{4}{84}, \; p(0, 1) = \frac{18}{84}, \; p(0, 2) = \frac{12}{84}, \; p(0, 3) = \frac{1}{84}, \; p(1, 0) = \frac{12}{84},$$
$$p(1, 1) = \frac{24}{84}, \; p(1, 2) = \frac{6}{84}, \; p(2, 0) = \frac{4}{84}, \; p(2, 1) = \frac{3}{84}. \tag{5.58}$$

Find the mean, variance and standard deviation of X and Y. Also find the covariance and the correlation coefficient.

Solution

The means are found using the formulae (5.53) for the mean:

$$\mu_X = 0p(0,0) + 0p(0,1) + 0p(0,2) + 0p(0,3) + 1p(1,0) + 1p(1,1) + 1p(1,2)$$
$$+ 2p(2,0) + 2p(2,1)$$
$$= 0 + 0 + 0 + 0 + 1\left(\frac{12}{84}\right) + 1\left(\frac{24}{84}\right) + 1\left(\frac{6}{84}\right) + 2\left(\frac{4}{84}\right) + 2\left(\frac{3}{84}\right) = \frac{56}{84}.$$
$$= 0.67$$

$$\mu_Y = 0p(0,0) + 0p(1,0) + 0p(2,0) + 1p(0,1) + 1p(1,1) + 1p(2,1)$$
$$+ 2p(0,2) + 2p(1,2) + 3p(0,3)$$
$$= 0 + 0 + 0 + 1\left(\frac{18}{84}\right) + 1\left(\frac{24}{84}\right) + 1\left(\frac{3}{84}\right) + 2\left(\frac{12}{84}\right) + 2\left(\frac{6}{84}\right) + 3\left(\frac{1}{84}\right)$$
$$= 1.$$

The variances are calculated using the variance formulae of (5.53):

$$\sigma_X^2 = \left(0 - \frac{56}{84}\right)^2 p(0,0) + \left(0 - \frac{56}{84}\right)^2 p(0,1) + \left(0 - \frac{56}{84}\right)^2 p(0,2) + \left(0 - \frac{56}{84}\right)^2 p(0,3)$$
$$+ \left(1 - \frac{56}{84}\right)^2 p(1,0) + \left(1 - \frac{56}{84}\right)^2 p(1,1) + \left(1 - \frac{56}{84}\right)^2 p(1,2)$$
$$+ \left(2 - \frac{56}{84}\right)^2 p(2,0) + \left(2 - \frac{56}{84}\right)^2 p(2,1)$$
$$= \left(0 - \frac{56}{84}\right)^2 \left(\frac{4}{84}\right) + \left(0 - \frac{56}{84}\right)^2 \left(\frac{18}{84}\right) + \left(0 - \frac{56}{84}\right)^2 \left(\frac{12}{84}\right) + \left(0 - \frac{56}{84}\right)^2 \left(\frac{1}{84}\right)$$
$$+ \left(1 - \frac{56}{84}\right)^2 \left(\frac{12}{84}\right) + \left(1 - \frac{56}{84}\right)^2 \left(\frac{24}{84}\right) + \left(1 - \frac{56}{84}\right)^2 \left(\frac{6}{84}\right)$$
$$+ \left(2 - \frac{56}{84}\right)^2 \left(\frac{4}{84}\right) + \left(2 - \frac{56}{84}\right)^2 \left(\frac{3}{84}\right)$$
$$= \frac{7}{18}.$$

$$\sigma_Y^2 = (0 - 1)^2 p(0,0) + (0 - 1)^2 p(1,0) + (0 - 1)^2 p(2,0)$$
$$+ (1 - 1)^2 p(0,1) + (1 - 1)^2 p(1,1) + (1 - 1)^2 p(2,1)$$
$$+ (2 - 1)^2 p(0,2) + (2 - 1)^2 p(1,2) + (3 - 1)^2 p(0,3)$$
$$= (0 - 1)^2 \left(\frac{4}{84}\right) + (0 - 1)^2 \left(\frac{12}{84}\right) + (0 - 1)^2 \left(\frac{4}{84}\right)$$
$$+ 0 + 0 + 0 + (2 - 1)^2 \left(\frac{12}{84}\right) + (2 - 1)^2 \left(\frac{6}{84}\right) + (3 - 1)^2 \left(\frac{1}{84}\right)$$
$$= \frac{1}{2}.$$

The standard deviations are

$$\sigma_X = 0.6236, \quad \text{and} \quad \sigma_Y = 0.7071.$$

The covariance is found using formula (5.55):

$$
\begin{aligned}
\sigma_{XY} &= \left(0 - \frac{56}{84}\right)(0-1)p(0,0) + \left(0 - \frac{56}{84}\right)(1-1)p(0,1) \\
&\quad + \left(0 - \frac{56}{84}\right)(2-1)p(0,2) + \left(0 - \frac{56}{84}\right)(3-1)p(0,3) \\
&\quad + \left(1 - \frac{56}{84}\right)(0-1)p(1,0) + \left(1 - \frac{56}{84}\right)(1-1)p(1,1) + \left(1 - \frac{56}{84}\right)(2-1)p(1,2) \\
&\quad + \left(2 - \frac{56}{84}\right)(0-1)p(2,0) + \left(2 - \frac{56}{84}\right)(1-1)p(2,1) \\
&= \left(0 - \frac{56}{84}\right)(0-1)\left(\frac{4}{84}\right) + \left(0 - \frac{56}{84}\right)(1-1)\left(\frac{18}{84}\right) \\
&\quad + \left(0 - \frac{56}{84}\right)(2-1)\left(\frac{12}{84}\right) + \left(0 - \frac{56}{84}\right)(3-1)\left(\frac{1}{84}\right) \\
&\quad + \left(1 - \frac{56}{84}\right)(0-1)\left(\frac{12}{84}\right) + \left(1 - \frac{56}{84}\right)(1-1)\left(\frac{24}{84}\right) + \left(1 - \frac{56}{84}\right)(2-1)\left(\frac{6}{84}\right) \\
&\quad + \left(2 - \frac{56}{84}\right)(0-1)\left(\frac{4}{84}\right) + \left(2 - \frac{56}{84}\right)(1-1)\left(\frac{3}{84}\right) \\
&= -\frac{1}{6} = -0.1667.
\end{aligned}
$$

The correlation coefficient ρ is found using formula (5.57):

$$\rho = \frac{\sigma_{XY}}{\sigma_X \sigma_Y} = \frac{-0.1667}{(0.6236)(0.7071)} = -0.3780$$

Example 5.4.34 (Variance and standard deviation of a continuous joint distribution) Determine the mean, variance, standard deviation, covariance and the correlation function of the joint probability density given in example 5.3.24, which we restate here:

$$
p(x, y) = \begin{cases} e^{-2x}e^{-\frac{y}{2}} & \text{for } 0 \le x \le \infty,\ 0 \le y \le \infty \\ 0 & \text{otherwise.} \end{cases}
$$

Solution
The means are calculated using the formulae for the mean in Eq. (5.54):

$$\mu_X = \int_{-\infty}^{\infty} \int_{\infty}^{\infty} xe^{-2x} e^{-\frac{y}{2}} \, dy \, dx = \int_{0}^{\infty} \int_{0}^{\infty} xe^{-2x} e^{-\frac{y}{2}} \, dy \, dx$$

$$= \int_{0}^{\infty} xe^{-2x} \left[-2e^{-\frac{y}{2}} \right]_{0}^{\infty} dx = -2 \left[e^{-2x} \left(-0.25 - 0.5x \right) \right]_{0}^{\infty}$$

$$= 0.5.$$

$$\mu_Y = \int_{0}^{\infty} \int_{0}^{\infty} ye^{-2x} e^{-\frac{y}{2}} \, dy \, dx$$

$$= \int_{0}^{\infty} ye^{-\frac{y}{2}} \left[-\frac{1}{2} e^{-2x} \right]_{0}^{\infty} dx = -\frac{1}{2} \left[e^{-\frac{y}{2}} \left(-4 - 2y \right) \right]_{0}^{\infty}$$

$$= 2.$$

The variances are calculated using the formulae for the variance in Eq. (5.54):

$$\sigma_X^2 = \int_{0}^{\infty} \int_{0}^{\infty} (x - 0.5)^2 e^{-2x} e^{-\frac{y}{2}} \, dy \, dx$$

$$= \int_{0}^{\infty} (x - 0.5)^2 e^{-2x} \left[-2e^{-\frac{y}{2}} \right]_{0}^{\infty} dx = -2 \left[e^{-2x} \left(-0.125 - 0.5x^2 \right) \right]_{0}^{\infty}$$

$$= 0.25.$$

$$\sigma_Y^2 = \int_{0}^{\infty} \int_{0}^{\infty} (y - 2)^2 e^{-2x} e^{-\frac{y}{2}} \, dy \, dx$$

$$= \int_{0}^{\infty} (y - 2)^2 e^{-\frac{y}{2}} \left[-\frac{1}{2} e^{-2x} \right]_{0}^{\infty} dx = -\frac{1}{2} \left[-2e^{-\frac{y}{2}} \left(4 + y^2 \right) \right]_{0}^{\infty}$$

$$= 4.$$

The standard deviations are

$$\sigma_X = 0.5 \quad \text{and} \quad \sigma_Y = 2.$$

The covariance is calculated using Eq. (5.56):

$$\sigma_{XY} = \int_{0}^{\infty} \int_{0}^{\infty} (x - 0.5)(y - 2)e^{-2x} e^{-\frac{y}{2}} \, dy \, dx = 0.$$

Since the covariance σ_{XY} is 0, the correlation coefficient is 0. This shows that the random variables X and Y are uncorrelated (statistically independent). This is to be expected, since, as mentioned earlier in this section, whenever a joint probability density can be written as a product of functions, each depending on only one random variable, the random variables are statistically independent.

Conditional Mean and Variance

We may also define the conditional mean and variance of X and Y by using the conditional probability densities $p(y|x)$ and $p(x|y)$ defined in Eq. (5.30) and Eq. (5.31), respectively:

Conditional mean of Y	$\mu_{Y|X} = \int_{-\infty}^{\infty} y\, p(y|x)\, dy$
Conditional mean of X	$\mu_{X|Y} = \int_{-\infty}^{\infty} x\, p(x|y)\, dx$

The formulae apply both to the continuous and the discrete case, but for the continuous case, $X = x$ should be interpreted as $x < X \leq x + dx$, similarly for $Y = y$.

Other Statistical Measures

Aside from the mean, the most relevant, there are other measures of *central tendency*, and aside from the variance, there are other measures of dispersion (spread). As a measure of central tendency the *mode* or the *median* is sometimes used. The *mode* is that value of x, call it x_m, that occurs most often. From this it follows that the probability $p(x_m)$ is a maximum. Some distributions may have more than one mode. The *median* is the value of x for which $P(X \leq x) = P(X \geq x) = \frac{1}{2}$. Another measure of dispersion is the *range*, defined as the difference between the largest and smallest value of a data set.

5.5 Simple Random Walk

The simple random walk concerns finding the probability $P(m)$ that after taking a total of $N = n_1 + n_2$ steps, with $n_1 = $ steps to the right and $n_2 = $ steps to the left, along a straight line, a particle (or person) ends up $m = n_1 - n_2$ steps from the origin. Let $p = $ *probability of a step to the right* and $q = 1 - p = $ *probability of a step to the left*. It is assumed that each step is independent of any other. We first ask for the probability of a specific sequence of n_1 steps to the right and n_2 steps to the left. To find this probability we use the generalised multiplication rule for the case of independent events, Eq. (5.6). In words, formula (5.6) states that the probability that events $A_1, A_2, \ldots, A_N$ occur together is the product of their probabilities. Using this, the probability for a specific sequence of n_1 steps to the right and n_2 steps to the left is

$$\underbrace{pp\cdots p}_{n_1}\,\underbrace{qq\cdots q}_{n_2} = p^{n_1} q^{n_2}. \tag{5.59}$$

This specific sequence can occur a number of different ways, and this number is equal to the number of ways of arranging N objects n_1 ways in any order, i.e., the number of ways of combining N objects n_1 ways (or, equivalently, N objects n_2 ways). We saw in subsection 5.2.5 that this number of combinations is

$$_N C_{n_1} = \binom{N}{n_1} = \frac{N!}{(N - n_1)! n_1!} = \frac{N!}{n_1! n_2!}.$$

The total probability $P(n_1, n_2)$ of taking n_1 steps to the right and n_2 steps to the left in any sequence is just the probability of one sequence summed $_N C_n$ times, i.e.,

$$P(n_1, n_2) = \binom{N}{n_1} p^{n_1} q^{n_2} = \frac{N!}{n_2! n_1!} p^{n_1} q^{n_2}.$$

Since $m = n_1 - n_2$, $P(m) = P(n_1, n_2)$ so that the probability $P(m)$ of ending up m steps from the origin after N steps is

$$P(m) = \binom{N}{n_1} p^{n_1} q^{n_2} = \frac{N!}{n_2! n_1!} p^{n_1} q^{n_2}.$$

5.6 Binomial, Poisson, Hypergeometric and Gaussian Distributions

The distribution of data depends on numerous underlying factors and different random experiments produce a great variety of different distributions. But, commonly encountered random experiments of interest produce data distributions which can be described by a small set of distributions. We shall consider the most important of these: the binomial, Poisson, hypergeometric and Gaussian distributions. The binomial and hypergeometric distributions are discrete finite distributions, while the Poisson distribution is a discrete infinite distribution. The Gaussian distribution is continuous, and is perhaps the most important, not least because of its mathematical simplicity. It is also the most relevant for distributions of measured values.

5.6.1 Binomial Distribution

The *binomial distribution* or *Bernoulli distribution*, named after the Swiss mathematician Jacques Bernoulli,[6] is a discrete distribution that occurs in games of chance (e.g., tossing a coin), quality inspection (e.g., counting the number of defective items),

---

[6]The Bernoulli distribution is so named because J. Bernoulli was the first to study problems leading to this distribution. In his 1713 book *Ars Conjectandi (The Art of Conjuring)* he included a treatment of the problem of independent trials having two equally probable outcomes. He tried to show that

opinion polls (e.g., number of people who like a particular brand of coffee), medicine (e.g., number of people that respond favourably to a drug under test) etc..

Generally, the binomial distribution applies to any set of trials with only two independent, equally probable outcomes. The outcomes are invariably labeled 'success' or 'failure'. The outcomes are associated with the random variable X, which has two values: $X = 1 = a$ *success*, and $X = 0 = a$ *failure*. All successful outcomes $X = 1$ have the same probability $p(1) = p$, so that failures $X = 0$ must have the probability $p(0) = 1 - p = q$. The interest is to find the probability $p(i)$ of i successes in n trials. This probability is given by

Binomial distribution or **Bernoulli distribution**	$$P(X = i) = p(i) = \binom{n}{i} p^i q^{n-i},$$ $$i = 0, 1, 2, \ldots, n$$	(5.60)

Since the distribution contains the binomial coefficients we see why it is called the binomial distribution. It is not difficult to show how this formula is obtained. The proof follows:

Consider the following sequence of i successes and $n - i$ failures:

$$\underbrace{sss \ldots s}_{i} \quad \underbrace{fff \ldots f}_{n-i}$$

Since the probabilities are independent, the probability that all of these successes and failures occur is just the product of the probabilities of each outcome as given in Eq. (5.6)

$$p^i (1 - p)^{n-i}. \tag{5.61}$$

But there are a number of different sequences with i successes and $n - i$ failures in different positions in the sequence. Note that for each different sequence, the order of the s's does not matter, i.e., interchanging s's among themselves does not change the sequence. Similarly, the order of the f's does not matter. In this case the number of different sequences is given by the formula for combinations, Eq. (5.9):

$$\binom{n}{i}. \tag{5.62}$$

It follows that the total probability $p(i)$ for i successes in n trials is the product of Eqs. (5.62) and (5.61):

$$p(i) = \binom{n}{i} p^i (1 - p)^{n-i} = \binom{n}{i} p^i q^{n-i}, \tag{5.63}$$

---

for a large enough number of trials the relative frequency of successful outcomes would approach the probability for a successful outcome, but failed.

which completes the proof. By using the binomial theorem we can easily show that the binomial distribution satisfies the fundamental requirement of a probability function, condition 2 of Eq. (5.13):

$$\sum_{i=0}^{n} p(i) = \sum_{i=0}^{n} \binom{n}{i} p^i q^{n-i} = (p+q)^n = [p + (1-p)]^n = 1.$$

Since p is necessarily positive, condition 1 of Eq. (5.13) is automatically satisfied. The binomial distribution has the following important properties:

Properties of the Binomial Distribution	
Mean	$\mu = np$
Variance	$\sigma^2 = npq$
Standard deviation	$\sigma = \sqrt{npq}$

Proof that mean = $\mu = np$: By substituting the binomial distribution for $p(x_i)$ in the definition of the mean, Eq. (5.39), we obtain an expression for the mean $\langle X^k = i^k \rangle$ of a Bernoulli random variable X:

$$\langle X^k = i^k \rangle = \langle i^k \rangle = \sum_{i=0}^{\infty} i^k \binom{n}{i} p^i q^{n-i} = \sum_{i=1}^{\infty} i^k \binom{n}{i} p^i q^{n-i},$$

noting that the $i = 0$ term is zero. Using the identity

$$i \binom{n}{i} = n \binom{n-1}{i-1}$$

we get

$$\langle X^k = i^k \rangle = \sum_{i=1}^{\infty} i^k \frac{n}{i} \binom{n-1}{i-1} p^i q^{n-i}$$

$$= np \sum_{i=1}^{\infty} i^{k-1} \binom{n-1}{i-1} p^{i-1} q^{n-i}$$

$$= np \sum_{j=0}^{\infty} (j+1)^{k-1} \binom{n-1}{j} p^j q^{(n-1)-j}, \quad \text{by setting } j = i - 1$$

$$= np \langle (Y+1)^{k-1} = (j+1)^{k-1} \rangle, \tag{5.64}$$

where Y is a Bernoulli random variable with parameters $(n-1, p)$. Setting $k = 1$ in Eq. (5.64) gives the mean of a binomial distribution

$$\text{mean} = \mu = \langle X = i \rangle = np\langle (Y+1)^0 = (j+1)^0 \rangle = np\langle 1 \rangle = np, \qquad (5.65)$$

which completes the proof.

Proof that variance $= \sigma^2 = npq$: This time, setting $k = 2$ in Eq. (5.64) gives

$$\langle X^2 = i^2 \rangle = np\langle (Y+1)^1 = (j+1)^1 \rangle = np\,(\langle Y \rangle + 1)$$
$$= np[(n-1)p + 1] = n^2 p^2 - np^2 + np, \qquad (5.66)$$

where the result $\langle Y \rangle = (n-1)p$ follows from Eq. (5.65), since Y is a Bernoulli random variable with parameters $(n-1, p)$. Substituting Eqs. (5.65) and (5.66) into the formula for the variance Eq. (5.50) gives

$$\sigma^2 = \langle X^2 \rangle - \langle X \rangle^2 = n^2 p^2 - np^2 + np - n^2 p^2 = np - np^2 = np(1-p) = npq,$$

which completes the proof.

Fig. 5.4 shows that a binomial distribution approaches a Gaussian distribution as the number of trails n increases.

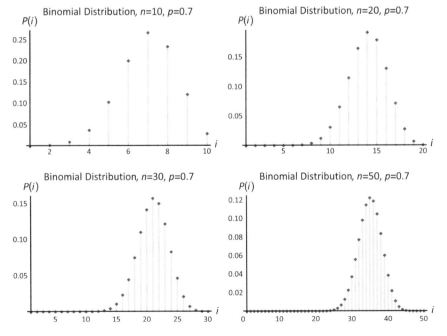

Fig. 5.4 The four plots show that the binomial distribution characterised by (n, p) approaches the Gaussian distribution (see Sect. 5.6.4) as n increases

Example 5.6.35 (Coin tossing. Application of the Binomial Distribution I)
Four coins are tossed. Determine the probability density for the number of heads obtained.

Solution

Let the random variable $X = number\ of\ heads$ (successes) $= \{0, 1, 2, 3, 4\}$. Since each trial has two equally probable outcomes, i.e, $p = \frac{1}{2}, q = 1 - \frac{1}{2} = \frac{1}{2}$, we can use the binomial distribution Eq. (5.60) with $n = 4$ and $p = \frac{1}{2}$:

$$P(X = 0) = \binom{4}{0}\left(\frac{1}{2}\right)^0\left(1 - \frac{1}{2}\right)^4 = \frac{1}{16}, \quad P(X = 1) = \binom{4}{1}\left(\frac{1}{2}\right)^1\left(1 - \frac{1}{2}\right)^3 = \frac{4}{16},$$

$$P(X = 2) = \binom{4}{2}\left(\frac{1}{2}\right)^2\left(1 - \frac{1}{2}\right)^2 = \frac{6}{16}, \quad P(X = 3) = \binom{4}{3}\left(\frac{1}{2}\right)^3\left(1 - \frac{1}{2}\right)^1 = \frac{4}{16}$$

$$P(X = 4) = \binom{4}{4}\left(\frac{1}{2}\right)^4\left(1 - \frac{1}{2}\right)^0 = \frac{1}{16}.$$

These, of course, are the values also obtained by counting desired outcomes and dividing by the total number of points, 16, of the sample space.

Example 5.6.36 (Defective fuses. Application of the Binomial Distribution II)
Electric fuses are sold in packets of 20. All fuses have an equal probability $p = 0.007$ of being defective. The probability of one fuse being defective is independent of the probability of another fuse being defective. A money-back guarantee is offered if more than one fuse in a packet is defective. What percentage of fuse packets are refunded?

Solution

The trial, which consists of testing a fuse, has two outcomes, defective or good, so we can use the binomial distribution Eq. (5.60) with $n = 20$ and $p = 0.007$. Let the random variable $X = number\ of\ defective\ fuses$. Packets with 0 or 1 defective fuse are not returned. The probabilities of these packets turning up is found as follows:

$$P(X = 0) = \binom{20}{0} 0.007^0 (1 - 0.007)^{20} = 0.86893,$$

$$P(X = 1) = \binom{20}{1} 0.007^1 (1 - 0.007)^{19} = 0.122508.$$

The total probability of a packet not being returned is $P(X = 0) + P(X = 1)$, from which it follows that the probability $P(\text{returned})$ of a packet being returned is

$$P(\text{returned}) = 1 - P(X = 0) - P(X = 1) = 0.0086.$$

We conclude that 0.86% of packets will be refunded.

5.6.2 The Poisson Distribution

As the name implies, the *Poisson distribution* (or *Poisson probability density*) was first derived in 1837 by the French mathematician Siméon Denis Poisson.[7] The Poisson distribution is the limit of the binomial distribution as $p \to 0$ and $n \to \infty$, while $\mu = np$ remains finite. It follows that the Poisson and binomial distributions are closely related. Indeed, for large n and small p, the Poisson distribution, preferred for calculation, serves as a good approximation to the binomial distribution. Essentially, the Poisson distribution is the generalisation of the binomial distribution to a discrete countable infinity of trials. As for a binomial distribution, the trials are ideally independent, but also like the binomial distribution, for a large population, the difference between results from independent or dependent trials is not too large. In this case, the Poisson distribution can serve as a reasonable approximation for trials that are not independent.

The Poisson distribution has a very wide area of application since it serves as an approximation to the binomial distribution when the probability p of successes is small while the number of trials n is large. Under these conditions the use of the Poisson distribution is preferred since calculations with the Poisson distribution are much easier. Further, the Poisson distribution best describes distributions which arise from natural processes where values may change at any instant of time. An important example from nuclear physics concerns particle counts in a fixed time interval (e.g., α or β- particles) or radiation (e.g., γ or X-rays) emitted by a radioactive substance (e.g., nickel oxide).

The Poisson distribution (Fig. 5.5) is given by

Poisson Distribution	$P(X = i) = p(i) = \dfrac{\mu^i}{i!} e^{-\mu}, \quad i = 0, 1, 2, \dots$ (5.67)

---

[7]The Poisson distribution was first presented in Poisson's 1837 book *Recherches sur la probabilité des judgements en matière criminelle et en matière civile* (*Investigations into the Probability of Verdicts in Criminal and Civil Matters*).

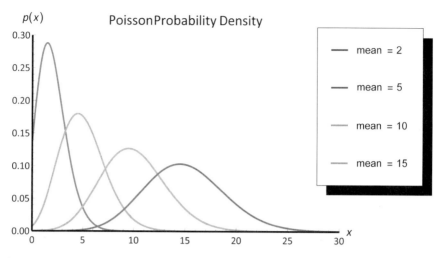

Fig. 5.5 The plots show the Poisson distribution for four values of the mean $\mu = 2, 5, 10$ and 15

In Eq. (5.67) μ is the mean and i represents the allowed values of the random variable X. That the distribution $p(i)$ satisfies the fundamental condition for a probability density, condition 2 of Eq. (5.13), is easily shown by summing $p(i)$ from $i = 0$ to ∞:

$$\sum_{i=0}^{\infty} p(i) = e^{-\mu} \sum_{i=0}^{\infty} \frac{\mu^i}{i!} = e^{-\mu} e^{\mu} = 1,$$

since the series is just the power series of the exponential function e^{μ}.

The Poisson distribution Eq. (5.67) can be derived from the binomial distribution in the following way: Begin with the binomial distribution, Eq. (5.60),

$$p(i) = \binom{n}{i} p^i q^{n-i} = \frac{n!}{(n-i)!\, i!} p^i (1-p)^{n-i}.$$

Introduce the parameter $\lambda = np$ and substitute $p = \frac{\lambda}{n}$:

$$p(i) = \frac{n!}{(n-i)!\, i!} \frac{\lambda^i}{n^i} \left(1 - \frac{\lambda}{n}\right)^{n-i}$$

$$= \frac{n(n-1)(n-2)\ldots(n-i+1)}{n^i} \frac{\lambda^i}{i!} \frac{(1-\lambda/n)^n}{(1-\lambda/n)^i} \tag{5.68}$$

As mentioned above, the Poisson distribution follows by taking the limit $p \to 0$ and $n \to \infty$, or equivalently, by taking the limit $\lambda \to 0$ and $n \to \infty$. Taking this limit, we get

$$\lim_{\lambda \to 0, n \to \infty} \left(1 - \frac{\lambda}{n}\right)^n \to e^{-\lambda}, \quad \lim_{n \to \infty} \frac{n(n-1)(n-2)\ldots(n-i+1)}{n^i} = 1,$$

$$\lim_{\lambda \to 0, n \to \infty} \left(1 - \frac{\lambda}{n}\right)^i \to 1$$

Substituting these results into Eq. (5.68) gives the Poisson distribution

$$P(X = i) = p(i) = \frac{\lambda^i}{i!} e^{-\lambda} = \frac{\mu^i}{i!} e^{-\mu},$$

where we have set $\lambda = \mu$ to get the last term.. This completes the derivation. The Poisson distribution has the following important properties:

Properties of the Poisson Distribution	
Mean	μ
Variance	$\sigma^2 = \mu$
Standard deviation	$\sigma = \sqrt{\mu}$

Proof that the mean $= \mu$: Substituting the Poisson distribution, Eq. (5.67), into the formula (5.39) for the mean gives

$$\text{Mean} = \langle X = i \rangle = \sum_{i=1}^{\infty} ip(i) = \sum_{i=1}^{\infty} i \frac{\mu^i}{i!} e^{-\mu}$$

$$= \mu e^{-\mu} \sum_{i=1}^{\infty} \frac{\mu^{i-1}}{(i-1)!}, \qquad \text{since the } i = 0 \text{ term is } 0$$

$$= \mu e^{-\mu} \sum_{j=0}^{\infty} \frac{\mu^j}{j!}, \qquad \text{by setting } j = i - 1$$

$$= \mu e^{-\mu} e^{\mu}, \qquad \text{since} \quad \sum_{i=0}^{\infty} \frac{\mu^j}{j!} = e^{\mu}$$

$$= \mu, \tag{5.69}$$

which completes the proof.

Proof that $\sigma^2 = \mu$: We begin by finding the mean of X^2:

$$\langle X^2 = i^2 \rangle = \sum_{i=1}^{\infty} i^2 \frac{\mu^i}{i!} e^{-\mu}$$

$$= e^{-\mu} \mu \sum_{i=1}^{\infty} i \frac{\mu^{i-1}}{(i-1)!} e^{-\mu}, \qquad \text{since the } i = 0 \text{ term is } 0$$

$$= e^{-\mu} \mu \sum_{j=0}^{\infty} (j+1) \frac{\mu^j}{j!}, \qquad \text{by setting } j = i - 1$$

$$= e^{-\mu} \mu \left[\sum_{i=0}^{\infty} j \frac{\mu^j}{j!} + \sum_{j=0}^{\infty} \frac{\mu^j}{j!} \right]$$

$$= e^{-\mu} e^{\mu} \mu (\mu + 1), \text{ since } \sum_{i=0}^{\infty} j \frac{\mu^j}{j!} = \mu \sum_{j=1}^{\infty} \frac{\mu^{j-1}}{(j-1)!} = \mu e^{\mu}, \text{ and } \sum_{j=0}^{\infty} \frac{\mu^j}{j!} = e^{\mu}$$

$$= \mu(\mu + 1)$$

$$= \mu^2 + \mu \tag{5.70}$$

Substituting Eqs. (5.69) and (5.70) into the formula for the variance Eq. (5.50) gives

$$\text{Variance} = \sigma^2 = \langle X^2 \rangle - \langle X \rangle^2 = \mu^2 + \mu - \mu^2 = \mu,$$

which completes the proof.

Example 5.6.37 (Faulty fuses. Application of the Poisson distribution I)
Consider again Example 5.6.36 . This time, calculate the probability that 0,1,2 and 5 fuses out of every 100 manufactured are faulty using the Poisson distribution with $\mu = np = 100(0.007) = 0.7$. Also calculate these probabilities using the binomial distribution for comparison.
Solution
Let $PP(i)$ represent the Poisson probabilities, and let $PB(i)$ represent the binomial probabilities. For the values $i = 0, 1, 2$ the probabilities are given by:

$$PB(0) = \binom{100}{0} (0.007)^0 (0.993)^{100} = 0.495364$$

$$PP(0) = e^{-0.7} \frac{0.7^0}{0!} = 0.496585$$

$$PB(1) = \binom{100}{1} (0.007)^1 (0.993)^{99} = 0.3492$$

$$PP(1) = e^{-0.7} \frac{0.7^1}{1!} = 0.34761$$

$$PB(2) = \binom{100}{2} (0.007)^2 (0.993)^{98} = 0.121851$$

$$PP(2) = e^{-0.7} \frac{0.7^2}{2!} = 0.121663$$

$$PB(5) = \binom{100}{5} (0.007)^5 (0.993)^{95} = 0.00064922$$

$$PP(5) = e^{-0.7} \frac{0.7^5}{5!} = 0.000695509$$

We see that the Poisson distribution probabilities are a good approximation to the binomial ones, as might be expected given that $n = 100$ is large and $p = 0.007$ is small, i.e., the criterion for the Poisson distribution to be a good approximation to the binomial distribution is satisfied.

Example 5.6.38 (Radioactive counting experiment. Application of the Poisson distribution II)

Consider an experiment in which the number of alpha particles emitted per second by 1 g of a radioactive substance are counted. Given that the average number of alpha particle counts per second is 3.4, calculate the probability $P(\leq 2)$ that no more than 2 alpha particles are counted in a 1 second interval.

Solution

Noting that $\mu = 3.4$ we get

$$P(0) = e^{-3.4} \frac{3.4^0}{0!} = 0.0333733$$

$$P(1) = e^{-3.4} \frac{3.4^1}{1!} = 0.113469$$

$$P(2) = e^{-3.4} \frac{3.4^2}{2!} = 0.192898.$$

Then

$$P(\leq 2) = P(0) + P(1) + P(2) = 0.33974.$$

5.6.3 The Hypergeometric Distribution

Drawing objects from a set of things (e.g., cards from a deck of cards) and replacing each object before the next object is picked guarantees independence of the trials.

In this case, the binomial distribution can be used whenever there are two outcomes. Suppose a bag contains N balls, M of which are red and $N - M$ are blue, then

$$\text{probability of a red ball} = p = \frac{M}{N}$$

$$\text{probability of a blue ball} = q = 1 - p = 1 - \frac{M}{N},$$

and Eq. (5.60) gives the probability $p(i)$ of picking i red balls in n trials:

$$p(i) = \binom{n}{i} p^i q^{n-i} = \binom{n}{i} \left(\frac{M}{N}\right)^i \left(1 - \frac{M}{N}\right)^{n-i}, \quad i = 0, 1, 2 \ldots$$

Suppose now the balls are picked without replacement. We again ask for the probability $p(i)$ of picking i red balls in n trials. The trials are no longer independent so that we cannot use the binomial distribution. Instead, we will determine this probability by counting the number of ways of getting i red balls in n trials. As usual, we will be greatly aided in counting these ways and sample points by the formula for combinations. The number of ways of choosing i red balls from the M red balls in the bag is, noting that $i \le M$,

$$\text{Number of ways of choosing } i \text{ red balls from } M \text{ red balls} = \binom{M}{i}.$$

Corresponding to each of these ways there is a number of ways of choosing $n - i$ blue balls in the remaining trials. This number is given by

$$\begin{array}{l} \text{For each way of choosing } i \text{ balls from } n \text{ trials,} \\ \text{the number of ways of choosing } (n - i) \text{ blue balls} \end{array} = \binom{N - M}{n - i}.$$

It follows that the total number of ways of choosing i red balls from n trials is just the product of these combinations:

$$\text{Total number of ways of choosing } i \text{ red balls from } n \text{ trials} = \binom{M}{i} \binom{N - M}{n - i}$$

The number of sample points is given by the number of ways of arranging the total number of balls N among n trials, noting that $n \le N$:

$$\text{Number of sample points} = \binom{N}{n}.$$

The probability p of picking i red balls in n trials without replacement is therefore given by

$$p(i) = \frac{\text{Number of ways of picking } i \text{ red balls in } n \text{ trials}}{\text{Number of sample points}} = \frac{\binom{M}{i}\binom{N-M}{n-i}}{\binom{N}{n}}.$$

(5.71)

Though we have derived this probability distribution for a specific case, it is general and describes trials with two outcomes when there is no replacement after each trial so that the trials are not independent. Contrast this with the binomial distribution which describes independent trials (replacement after each trial) with two outcomes. The probability function (5.71) is called the *hypergeometric distribution*.[8] We note that for large N, M and $N - M$ compared to n it does not matter whether or not there is replacement after each trial and the hypergeometric distribution can be approximated by the often simpler to use binomial distribution.

5.6.4 The Gaussian Distribution

The previous probability distribution functions we considered are all discrete. The *Gaussian distribution*, also called the *normal distribution* or *Gaussian probability density*, on the other hand, is continuous. It is, perhaps, the most important distribution since it describes the distributions resulting from a very wide variety of random processes or random experiments. Further, under suitable conditions, the Gaussian distribution approximates non-Gaussian distributions. For example, for a large number n of trials, the binomial distribution approaches the Gaussian distribution (Fig. 5.4), so that, in such cases, the often mathematically simpler Gaussian distribution serves as a good approximation.

The Gaussian distribution was actually introduced by the French mathematician Abraham de Moivre in 1733, who developed it as an approximation to the binomial distribution for a large number of trials. De Moivre was concerned with calculating probabilities in games of chance. It was brought to prominence in 1809 by the German mathematician Karl Friedrich Gauss, who applied the distribution to astronomical problems. Since this time it has become known as the Gaussian distribution. The Gaussian distribution described so many data sets that in the latter part of the 19th century the British statistician Karl Pearson coined the name 'normal distribution', a name which, like the name 'Gaussian distribution,' has stuck to this day.

---

[8]It is so called because its moment generating function (defined in Eq. (5.52)) can be expressed as a hypergeometric function.

An example of a random experiment that gives rise to a Gaussian distribution is the measurement of a physical quantity such as the height of a person or the velocity of an object. This example is particularly relevant to the aim of earlier chapters since the Gaussian distribution plays an essential role in the theory of errors. It is argued that the numerous perturbations that occur in each measurement push the result up or down with equal measure. This tends to reduce the error in the measurement and to produce the characteristic symmetric bell shape of the Gaussian distribution. We should keep in mind that though many measured quantities are continuous, such as the height of a person, each measured result is necessarily finite (to a fixed number of significant figures) as are the total number of measurements. For large amounts of data, fitting the finite measurements to a Gaussian distribution not only makes calculation and analysis much easier, it also provides a better approximation to the true mean - in a sense the continuous curve of the Gaussian distribution 'fills in' gaps between the finite measurements and the infinite number of measurements needed to establish the (unknowable) true mean, so providing a better estimate.

The Gaussian distribution is defined by

| **Gaussian Distribution** | $$P(X = x) = p(x) = \frac{1}{\sigma\sqrt{2\pi}} e^{-\frac{(x-\mu)^2}{2\sigma^2}}, \quad -\infty < x < \infty \quad (5.72)$$ |

It is straightforward to show that the probability density $p(x)$ satisfies the fundamental condition for a probability function, condition 2 of (5.14). To do this we first introduce the *standardised random variable Z* corresponding to the random variable X:

$$Z = \frac{X - \mu}{\sigma} \qquad (5.73)$$

If X has mean μ and variance σ^2, then Z has mean 0 and variance 1.

We begin the proof by substituting Z and $dx = \sigma dz$ into Eq. (5.72):

$$\frac{1}{\sigma\sqrt{2\pi}} \int_{-\infty}^{\infty} e^{-\frac{(x-\mu)^2}{2\sigma^2}} dx = \frac{1}{\sqrt{2\pi}} \int_{-\infty}^{\infty} e^{-\frac{z^2}{2}} dz. \qquad (5.74)$$

From tables

$$\int_{-\infty}^{\infty} e^{-az^2} dz = 2 \int_{0}^{\infty} e^{-az^2} dz = \sqrt{\frac{\pi}{a}}.$$

Substituting $a = \frac{1}{2}$ gives

$$\int_{-\infty}^{\infty} e^{-\frac{z^2}{2}} dz = \sqrt{2\pi}. \qquad (5.75)$$

Substituting Eq. (5.75) into Eq. (5.74) gives

$$\int_{-\infty}^{\infty} \frac{1}{\sigma\sqrt{2\pi}} e^{-\frac{(x-\mu)^2}{2\sigma^2}} \, dx = \frac{\sqrt{2\pi}}{\sqrt{2\pi}} = 1,$$

which completes the proof.

The corresponding distribution function is

$$F(x) = P(X \le x) = \frac{1}{\sigma\sqrt{2\pi}} \int_{-\infty}^{x} e^{-\frac{(v-\mu)^2}{2\sigma^2}} \, dv. \tag{5.76}$$

Substituting the standardised random variable $Z = z$ given in Eq. (5.73) into Eq. (5.72) gives the *standard Gaussian probability distribution* or *standard Gaussian probability density*,

$$p(z) = \frac{1}{\sqrt{2\pi}} e^{-\frac{z^2}{2}}, \tag{5.77}$$

while substituting it into Eq. (5.76) gives the *standard Gaussian distribution function*

$$F(z) = P(Z \le z) = \frac{1}{\sqrt{2\pi}} \int_{-\infty}^{z} e^{-\frac{u^2}{2}} \, du = \frac{1}{2} + \frac{1}{\sqrt{2\pi}} \int_{0}^{z} e^{-\frac{u^2}{2}} \, du \tag{5.78}$$

The distribution function $F(z)$ is closely related to the error function $\mathrm{erf}(z)$, defined by

$$\mathrm{erf}(z) = \frac{2}{\sqrt{\pi}} \int_{0}^{z} e^{-u^2} \, du,$$

so that $F(z)$ can be written in terms of $\mathrm{erf}(z)$:

$$F(z) = \frac{1}{2}\left[1 + \mathrm{erf}\left(\frac{z}{\sqrt{2}}\right)\right]$$

A useful relation for finding $F(z)$ for negative z is

$$F(-z) = 1 - F(z)$$

The following relations, which follow from the above definitions, are useful in calculations of probabilities such as $P(X \le a)$ or $P(a \le X \le b)$:

$$F(a) = P(X \le a) = P\left(\frac{X-\mu}{\sigma} \le \frac{a-\mu}{\sigma}\right) = F\left(\frac{a-\mu}{\sigma}\right) = F(A)$$

$$= \frac{1}{\sqrt{2\pi}} \int_{-\infty}^{A} e^{-\frac{z^2}{2}} \, dz \tag{5.79}$$

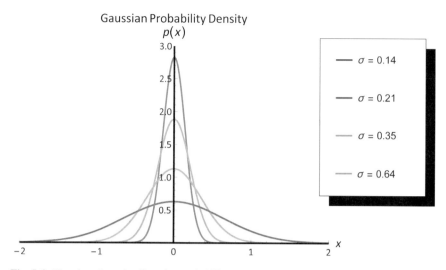

Fig. 5.6 The plots show the Gaussian probability density (Gaussian distribution) for mean $\mu = 0$ and four values of the standard deviation, $\sigma = 0.14, 0.21, 0.35$ and 0.64

$$P(a \leq X \leq b) = P\left(\frac{a - \mu}{\sigma} \leq \frac{X - \mu}{\sigma} \leq \frac{b - \mu}{\sigma}\right)$$

$$= F\left(\frac{b - \mu}{\sigma}\right) - F\left(\frac{a - \mu}{\sigma}\right) = F(B) - F(A)$$

$$= \frac{1}{\sqrt{2\pi}} \int_A^B e^{-\frac{z^2}{2}} \, dz \qquad (5.80)$$

Here are some properties of the Gaussian distribution (Fig. 5.6):

Properties of the Gaussian Distribution	
Mean	μ
Variance	σ^2
Standard deviation	σ

Proof that mean $= \mu$: Beginning with the definition of the mean, we have

$$\langle X \rangle = \frac{1}{\sigma\sqrt{2\pi}} \int_{-\infty}^{\infty} x e^{-\frac{(x-\mu)^2}{2\sigma^2}} \, dx.$$

Substitute the standardised variable Z of Eq. (5.73) together with $x = \mu + z\sigma$ and $dx = \sigma dz$ to get

$$\langle X \rangle = \frac{1}{\sqrt{2\pi}} \int_{-\infty}^{\infty} (\mu + z\sigma) e^{-\frac{z^2}{2}} \, dz$$

$$= \frac{\mu}{\sqrt{2\pi}} \int_{-\infty}^{\infty} e^{-\frac{z^2}{2}} \, dz + \frac{\sigma}{\sqrt{2\pi}} \int_{-\infty}^{\infty} z e^{-\frac{z^2}{2}} \, dz \qquad (5.81)$$

The first integral is the same as in Eq. (5.75) and equal to $\sqrt{2\pi}$. The second integral can by evaluated by noting that

$$z e^{-\frac{z^2}{2}} = -\frac{d}{dz} e^{-\frac{z^2}{2}}.$$

Using this result, the second integral becomes

$$-\int_{-\infty}^{\infty} \frac{d}{dz} e^{-\frac{z^2}{2}} \, dz = -\left[e^{-\frac{z^2}{2}} \right]_{-\infty}^{\infty} = 0$$

Substituting these values of the two integrals in Eq. (5.81) gives

$$\langle X \rangle = \frac{\mu}{\sqrt{2\pi}} (\sqrt{2\pi} + 0) = \mu, \qquad (5.82)$$

which completes the proof.

Proof that variance $= \sigma$: We first need the mean of X^2:

$$\langle X^2 \rangle = \frac{1}{\sigma\sqrt{2\pi}} \int_{-\infty}^{\infty} x^2 e^{-\frac{(x-\mu)^2}{2\sigma^2}} \, dx$$

Again, convert to standardised form using Z of Eq. (5.73):

$$\langle X^2 \rangle = \frac{1}{\sqrt{2\pi}} \int_{-\infty}^{\infty} (\mu + z\sigma)^2 e^{-\frac{z^2}{2}} \, dz$$

$$= \frac{1}{\sqrt{2\pi}} \int_{-\infty}^{\infty} (z^2 \sigma^2 + 2\mu\sigma z + \mu^2) e^{-\frac{z^2}{2}} \, dz$$

$$= \frac{\sigma^2}{\sqrt{2\pi}} \int_{-\infty}^{\infty} z^2 e^{-\frac{z^2}{2}} \, dz + \frac{2\mu\sigma}{\sqrt{2\pi}} \int_{-\infty}^{\infty} z e^{-\frac{z^2}{2}} \, dz + \frac{\mu^2}{\sqrt{2\pi}} \int_{-\infty}^{\infty} e^{-\frac{z^2}{2}} \, dz \, (5.83)$$

From above, the value of the second integral is 0, while the value of the third integral from Eq. (5.75) is $\sqrt{2\pi}$. The first integral can be evaluated using the following result from tables

$$\int_0^\infty z^{2n} e^{-az^2}\, dz = \frac{1 \cdot 3 \cdot 5 \cdots (2n-1)}{2^{n+1} a^n} \sqrt{\frac{\pi}{a}}$$

With $n = 1$ and $a = \frac{1}{2}$ and multiplying by 2 since the integration is from $-\infty$ to ∞ of a symmetric function, we get

$$\int_{-\infty}^\infty z^2 e^{-\frac{z^2}{2}}\, dz = \sqrt{2\pi}$$

Substitution of these values of the three integrals into Eq. (5.83) gives

$$\langle X^2 \rangle = \frac{\sigma^2}{\sqrt{2\pi}}\sqrt{2\pi} + 0 + \frac{\mu^2}{\sqrt{2\pi}}\sqrt{2\pi} = \sigma^2 + \mu^2. \tag{5.84}$$

Substituting Eqs. (5.82) and (5.84) into the formula for the variance Eq. (5.50) gives

$$\text{variance} = \langle X^2 \rangle - \langle X \rangle^2 = \sigma^2 + \mu^2 - \mu^2 = \sigma^2,$$

which completes the proof.

Example 5.6.39 (Calculation of probabilities of a Gaussian distributed random variable)

Let X be a random variable described by a Gaussian probability density with mean $\mu = 4$ and standard deviation $\sigma = 2$. Calculate the probability $P(3 \leq X \leq 8)$ that X has a value in the interval $3 \leq X \leq 8$ using (a) Eq. (5.15) with $p(x)$ given by Eq. (5.72), and (b) Eq. (5.80).

Solution

(a) Formula (5.15) and Eq. (5.72) gives

$$P(3 \leq X \leq 8) = \frac{1}{2\sqrt{2\pi}} \int_3^8 e^{\frac{-(x-4)^2}{2(4)}}\, dx = 0.668712,$$

which is the required probability.

(b) Formula (5.80) gives the same result as expected

$$P(3 \leq X \leq 8) = P\left(\frac{3-4}{2} \leq \frac{X-4}{2} \leq \frac{8-4}{2}\right)$$

$$= P\left(-\frac{1}{2} \leq Z \leq 2\right)$$

$$= \frac{1}{\sqrt{2\pi}} \int_{\frac{1}{2}}^2 e^{\frac{-z^2}{2}}\, dz = 0.668712 \tag{5.85}$$

Gaussian Approximation to the Binomial Distribution

That the binomial distribution can be approximated by the Gaussian distribution was proved by De Moivre in 1733 for the special case $p = \frac{1}{2}$, where p is the probability of success in the binomial distribution. In 1812 Laplace extended the proof for a general p. The result of the proof is expressed in what is now called the *De Moivre-Laplace limit theorem*:

Theorem 5.6.9 *The DeMoivre-Laplace limit theorem*

Theorem 5.6.9	*If X is a binomial random variable giving the number of successes in n binomial trials (trials with two independent outcomes), then its corresponding standardised random variable Z, given by* $$Z = \frac{X - \mu}{\sigma} = \frac{X - np}{\sqrt{npq}}, \text{ since } \mu = np \text{ and } \sigma = \sqrt{npq} \text{ for a}$$ *binomial distribution, approaches a standardised normal distribution as $n \to \infty$, i.e.,* $$\lim_{n \to \infty} P(a \leq Z \leq b) = F(b) - F(a) = \frac{1}{\sqrt{2\pi}} \int_a^b e^{-\frac{u^2}{2}} \, du,$$ *where p = probability of success and q = 1 − p probability of failure.*

From this theorem it follows that for large n (and p, q small) a standardised Gaussian distribution is a good approximation to a standardised binomial distribution.

Because we are approximating a discrete distribution with a continuous one, a correction called the continuity correction has to be made. Thus, to apply the approximation, we first make the correction

$$P(X = i) = P\left(i - \frac{1}{2} \leq X \leq i + \frac{1}{2}\right).$$

We see that there are two approximations to the binomial distribution: the Poisson approximation, which is good when n is large and p is small, and the Gaussian approximation, which is good if n is large and neither p nor q are too small. In practice, the Gaussian approximation is very good if both np and nq are greater than 5.

Example 5.6.39 (Gaussian approximation to the binomial distribution)
A coin is tossed 10 times. The random variable $X = $ *number of heads*. Calculate the probability $P(5 \leq X \leq 8)$ that either 5, 6, 7 or 8 heads are thrown using (a) a binomial distribution, and (b) a Gaussian approximation.
Solution
(a) Using the binomial distribution, the probabilities of getting 5, 6, 7 or 8 heads in 10 tosses of the coin are:

$$P(X = 5) = \binom{10}{5} 0.5^5 (1 - 0.5)^{10-5} = \frac{63}{256},$$

$$P(X = 6) = \binom{10}{6} 0.5^6 (1 - 0.5)^{10-6} = \frac{105}{512},$$

$$P(X = 7) = \binom{10}{7} 0.5^7 (1 - 0.5)^{10-7} = \frac{15}{128},$$

$$P(X = 8) = \binom{10}{8} 0.5^8 (1 - 0.5)^{10-8} = \frac{45}{1024}.$$

Therefore, the probability of getting 5, 6, 7 or 8 heads in 10 tosses of the coin is

$$P(5 \leq X \leq 8) = P(X = 5) + P(X = 6) + P(X = 7) + P(X = 8) = \frac{627}{1024} = 0.6123.$$

(b) To use the Gaussian approximation we first note that $\mu = np = 10(0.5) = 5$ and $\sigma = \sqrt{npq} = \sqrt{2.5}$. The probability we want is $P(5 \leq X \leq 8)$, but as we mentioned above, we must first apply the continuity correction, then transform to the standardised form.

The continuity correction gives

$$P(5 \leq X \leq 8) = P\left(5 - \frac{1}{2} \leq X \leq 8 + \frac{1}{2}\right) = P(4.5 \leq X \leq 8.5).$$

Standardising gives

$$P(4.5 \leq X \leq 8.5) = P\left(\frac{4.5 - 5}{\sqrt{2.5}} \leq \frac{X - 5}{\sqrt{2.5}} \leq \frac{8.5 - 5}{\sqrt{2.5}}\right) = P(-0.3162 \leq Z \leq 2.2135).$$

We can now use the standardised Gaussian approximation

$$P(4.5 \leq X \leq 8.5) = P(-0.3162 \leq Z \leq 2.2135) = \int_{-0.3162}^{2.2135} e^{-\frac{u^2}{2}} \, du = 0.6106.$$

Comparing with the binomial result $P(5 \leq X \leq 8) = 0.6123$, we see that the Gaussian approximation is quite good. Indeed, for our case $np = nq = 5$ satisfies the criterion for the Gaussian distribution to be a good approximation to the binomial distribution (Fig. 5.7).

Gaussian Approximation to the Poisson Distribution

Since the Gaussian distribution is related to the binomial distribution, and since the binomial distribution is related the Poisson distribution, then the Gaussian distribu-

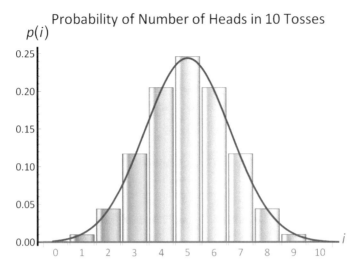

Fig. 5.7 A Gaussian probability density approximation to a binomial distribution for the number of heads obtained in 10 tosses of a coin

tion should be related to the Poisson distribution. This is indeed the case, and the relation is given by:

Theorem 5.6.9 *Relation of the Poisson and Gaussian distributions*

Theorem 5.6.9	*If X is a random variable with a Poisson distribution and corresponding standardised variable $Z = \frac{X-\mu}{\sigma}$, then* $$\lim_{\mu \to \infty} P\left(a \le \frac{X-\mu}{\sqrt{\mu}} \le b\right) = \frac{1}{\sqrt{2\pi}} \int_a^b e^{-\frac{u^2}{2^2}}\, du$$

This theorem allows the Poisson distribution to be approximated by the Gaussian distribution.

5.7 Three Important Theorems

We conclude with three important theorems that have played an important role in probability theory.

5.7.1 The Weak Law of Large Numbers

The *Weak Law of Large Numbers* was first derived by Swiss mathematician Jacques Bernoulli in his 1713 book *Ars Conjectandi* (*The Art of Conjuring*) for the special case of Bernoulli or binomial random variables (i.e., random variables produced by independent trials having two outcomes). It is stated as follows:

Theorem 5.7.9 *The Weak Law of Large Numbers*

Theorem 5.7.9	*Let $X_1, X_2, \ldots, X_n$ be a sequence of n independent random variables having identical distributions each with finite mean μ, then, for any $\epsilon > 0$,* $$\lim_{n \to \infty} P\left(\left\lvert \frac{X_1 + X_2 + \ldots + X_n}{n} - \mu \right\rvert \geq \epsilon\right) = 0$$

5.7.2 The Central Limit Theorem

A early version of the *central limit theorem* was proved by De Moivre around 1733 for the special case of Bernoulli random variables with $p = \frac{1}{2}$. Laplace also presented a special case of the central limit theorem in his 1812 book *Théorie analytique des probabilités* (*Analytic Theory of Probability*). Later, Laplace extended the theorem to arbitrary p. Laplace showed that the distribution of errors in large data samples gathered from astronomical observations was approximately Gaussian. Since error analysis is fundamental to all scientific experiment, Laplace's central limit theorem is regarded as a very important contribution to science.

The central limit theorem states that the sum of a large number of independent random variables is approximated by a value found from a Gaussian distribution. A more precise statement follows:

Theorem 5.7.9 *The Central Limit Theorem*

Theorem 5.7.9	*Let $X_1, X_2, \ldots, X_n$ be a sequence of n independent random variables having identical distributions each with mean μ and variance σ^2, then,* $$\lim_{n \to \infty} P\left(a \leq \frac{X_1 + X_2 + \ldots + X_n - n\mu}{\sigma\sqrt{n}} \leq b\right) = \frac{1}{\sqrt{2\pi}} \int_a^b e^{-\frac{x^2}{2^2}} \, dx$$

5.7.3 The Strong Law of Large Numbers

The Strong Law of Large Numbers was derived in 1909 for the special case of Bernoulli random variables by the French mathematician Emile Borel using the newly introduced *measure theory*. A general derivation was given later by A. N. Kolmogorov. The Strong Law states that the average of a sequence of independent random variables having the same distribution is certain (probability 1) to converge to the mean of the distribution. More precisely, the theorem may be stated thus:

Theorem 5.7.9 *The Strong Law of Large Numbers*

Theorem 5.7.9	*Let $X_1, X_2, \ldots, X_n$ be a sequence of n independent random variables having identical distributions each with finite mean μ, then, with probability 1,* $$\lim_{n \to \infty} \frac{X_1 + X_2 + \ldots + X_n}{n} = \mu$$

5.8 Problems

1. The outcome of spinning a roulette wheel is a number from 1 to 36 on a red or black background. Give two sample spaces that correspond to the outcomes of spinning a roulette wheel.

2. What is the probability of getting a 4, 5, or 6 of diamonds when drawing a card from a deck of 52 cards?

3. *An example of the use of Theorem 5.2.1.* Let event A_2 consist of the spades suite of a deck of 52 cards. Let event A_1 consist of the even numbered spades (we take jack $= 11$, queen $= 12$ and king $= 13$). The trial consists of drawing a card from the deck. Show that $P(A_2) \geq P(A_1)$ and determine $P(A_2) - P(A_1)$ by theorem 5.2.1 Confirm your answer for $P(A_2) - P(A_1)$ by direct counting of desired outcomes.

4. *An example of the use of Theorem 5.2.4.* Let A be composed of the events A_1 $=$ *king of hearts*, $A_2 =$ *king of diamonds*, $A_3 =$ *king of clubs* and $A_4 =$ *king of spades*. The trial consists of drawing one card from a deck of 52 cards. The event A of interest is drawing one of these cards. Find $P(A)$ both by direct counting of desired outcomes and by theorem 5.2.4, and hence confirm theorem 5.2.4.

5. *An example of the use of Theorem 5.2.5.* Let $A =$ *clubs suite of a deck of 52 cards* and $B =$ *cards numbered from 5 to 10*. The trial is drawing a card. Use theorem 5.2.5 to find the probability $P(A \cup B)$ of drawing a club or a card numbered 5 to 10. Confirm your answer by counting desired outcomes.

6. *An example of the use of Theorem 5.2.6.* Consider again events A and B and the trial of question 5. By calculating the three probabilities $P(A)$, $P(A \cap B)$ and $P(A \cap B')$ by direct counting, confirm theorem 5.2.6.

7. *An example of the use of Theorem 5.2.7.* Let A = *the hearts suite of a deck of 52 cards.* It consists of the mutually exclusive events: A_1 = *hearts ace to 3, A_2=hearts 4 to 7, A_3 = hearts 8 to 9* and A_4 = *10, jack, queen, king of hearts.* The trial consists of drawing a card. By determining the probability $P(A)$ that a heart is drawn and the probabilities of A occurring with A_1, A_2, A_3 and A_4, confirm theorem 5.2.7.

8. *Conditional Probability. Sampling with replacement.* Consider a bag containing 3 blue balls and 4 yellow balls. The trial is picking of a ball and is repeated twice. Determine the probability of picking a yellow ball, given that a yellow ball is already picked.

9. *The multiplication rule (I). Sampling with replacement.* Consider the ball-picking trials of question 8. Calculate the probability $P(A \cap B)$ of picking two yellow balls, with replacement, using the multiplication rule $P(A \cap B) = P(B|A)P(A)$, Eq. (5.2).

10. *Conditional Probability. Sampling without replacement.* Consider again question 8, but this time with the first picked ball not replaced. Again, calculate the probability of picking a yellow ball given that a yellow ball is already picked.

11. *The multiplication rule (II). Sampling without replacement.* Again, consider question 8 and ask for the probability $P(A \cap B)$ of picking 2 yellow balls from the bag using the multiplication rule $P(A \cap B) = P(B|A)P(A)$, Eq. (5.2), but this time without replacing the first picked ball.

12. *Multiplication Rule (III). Drawing cards (I).* What is the probability of drawing an 8 and a 9 from a deck of 52 cards with and without replacing the first chosen card?

13. *Multiplication Rule (IV). Drawing cards (II).* What is the probability of drawing three kings from a deck of 52 cards with and without replacing the first chosen card?

14. *Permutations.* How many permutations of 6 letters can be made from the letters of the word 'hippopotamus'?

15. *Combinations.* Consider again the arrangement of the letters of the word 'hippopotamus' in groups of 6 letters, but this time the different order of the same 6 letters does not matter. In other words, how many combinations of 6 letters can be made from the word 'hippopotamus'?

16. *Use of combinations and permutations in the calculation of probabilities.*
Four balls are picked, without replacement, from a bag containing 6 yellow balls and 4 green balls. Determine the probability of picking 2 yellow balls and 2 green balls using (a) permutations and (b) combinations.

17. *Probability function.* Consider the sample space

$$HHH, HHT, HTH, HTT, THH, THT, TTH, TTT$$

produced by tossing a coin three times. The random variable X = *number of heads* can take on values $X = 0, 1, 2, 3$. Write down the probability function for X.

18. *Distribution function.* Write down the distribution function corresponding to the probability function of problem 17.

19. *Probability density and distribution function.* Let the discrete random variable $X = product$ *of the two numbers when two dice are thrown.* Determine the probability density and distribution function for X.

20. *Continuous probability density.* The random variable X is described by the probability density

$$p(x) = \begin{cases} 0 \text{ for } x < -1 \\ \dfrac{1}{\sqrt{\pi}\,\mathrm{Erf}(1)} e^{-x^2} \text{ for } -1 \le x \le 1 \\ 0 \text{ for } x > 1. \end{cases}$$

Find (a) the distribution function, and (b) the probability $P(-0.5 \le X \le 0.5)$ using Eq. (5.15) and $F(b) - F(a)$ of Eq. (5.17).

21. *Discrete joint probabilities.* 3 balls are picked from a bag containing 3 orange balls, 4 yellow balls and 2 green balls. Let $X = number$ *of orange balls chosen*, and $Y = number$ *of yellow balls chosen*. (a) Define the joint probability density $P(X = x, Y = y) = p(x, y)$, (b) Determine the marginal probability density functions $P_x(X = x) = p_x(x)$ and $P_y(Y = y) = p_y(y)$, and (c) Draw a table to determine the various marginal probabilities and check the answer by finding the total probability.

22. *Continuous joint probability density.* The continuous joint probability density for two random variables X and Y is defined by

$$p(x, y) = \begin{cases} e^{-2x^2} e^{-\frac{y^2}{2}} \text{ for } -\infty \le x \le \infty, \ -\infty \le y \le \infty \\ 0 \qquad\qquad \text{otherwise.} \end{cases}$$

Determine (a) $P(X \ge 1, Y \le 1)$, (b) $P(X \le Y)$ and (c) $P(X \le a)$.

23. *Discrete conditional probability density.* Consider the discrete joint probability distribution

$$p(0, 0) = 0.1, \quad p(0, 1) = 0.4, \quad p(1, 0) = 0.2, \quad p(1, 1) = 0.3$$

for the random variables X and Y. Calculate the conditional probability density $p(x|y)$ given that $Y = y = 1$.

24. *Continuous conditional probability density.* Consider the continuous joint probability density

$$p(x, y) = \begin{cases} \dfrac{2}{21}(2x - 4)(3y - 5) \text{ for } 0 \le x \le 1, \ 0 \le y \le 1 \\ 0 \qquad\qquad\qquad \text{otherwise.} \end{cases}$$

Determine the conditional probability density $p(x|y)$ of getting X given that $Y = y$.

25. *Mean of a set of discrete quantities I.* A coin is tossed 3 times. The random variable $X = the\ number\ of\ heads\ after\ 3\ tosses$. Hence, $X = 0, 1, 2, 3$ with probabilities given by $P(X = 0) = \frac{1}{8}, P(X = 1) = \frac{3}{8}, P(X = 2) = \frac{3}{8}$ and $P(X = 3) = \frac{1}{8}$, where the number of sample points is 8. Find the mean.

26. *Mean of a set of discrete quantities II.* In an experiment, 10 measurements of the diameter of a copper wire were made, with the results given in millimeters: 0.51, 0.63, 0.59, 0.61, 0.54. 0.53, 0.59, 0.54, 0.63 and 0.54. Find the mean.

27. *Mean of a continuous quantity.* Find the expectation value of a random variable X given that it is described by the Gausssian probability density

$$p(x) = \frac{2}{\sqrt{\pi}} e^{-(2x-3)^2}.$$

28. *Mean of a function of a random variable.* The following are 6 measured values of one side of a cubical water tank in units of metres: 0.6986, 0.7634, 0.7286, 0.6629, 0.7041 and 0.6629. Let the random variable $X = the\ measured\ length\ of\ one\ side\ of\ the\ water\ tank$, and the function of the random variable $g(X) = the\ volume\ of\ the\ tank = X^3$. Find the means of X and $g(X)$.

29. *Variance and standard deviation of a set of discrete quantities.* Consider only the diamond suite of a deck of cards. Let the random variable $X = drawing\ an\ even\ numbered\ diamond$. Find the variance and standard deviation of X. Note that ace, jack, queen and king are counted as 1, 11, 12 and 13, respectively.

30. *Variance and standard deviation of a continuous quantity.* Find the variance and standard deviation of the random variable X described by the probability density

$$p(x) = \begin{cases} 4x^3 & \text{for } 0 \le x \le 1 \\ 0 & \text{otherwise.} \end{cases}$$

31. *Variance and standard deviation of a set of discrete joint distributions.* Consider again problem 21 in which 3 balls are picked from a bag containing 3 orange balls, 4 yellow balls and 2 green balls. Find the mean, variance and standard deviation of X and Y. Also find the covariance and the correlation coefficient.

32. *Variance and standard deviation of a set of continuous joint distributions.* Determine the mean, variance, standard deviation, covariance and the correlation coefficient of the following joint probability density:

$$p(x, y) = \begin{cases} \frac{3}{5}(2x^2 + 4xy) & \text{for } 0 \le x \le 3, \ 0 \le y \le 3 \\ 0 & \text{otherwise.} \end{cases}$$

33. *Coin tossing. Application of the binomial distribution I.* Three coins are tossed. Determine the probability density for the number of heads obtained using the binomial distribution.

34. *Defective fuses. Application of the binomial distribution II.* Electric fuses are sold in packets of 30. All fuses have an equal probability $p = 0.005$ of being defective The probability of one fuse being defective is independent of the probability of another fuse being defective. A money-back guarantee is offered if more than one fuse in a packet is defective. What percentage of fuse packets are refunded?

35. *Faulty fuses. Application of the poisson distribution I.* Consider problem 34. This time, calculate the probability that 0,1,2 and 5 fuses out of every 200 manufactured are faulty using the Poisson distribution with $\mu = np = 200(0.005) = 1$. Also calculate these probabilities using the binomial distribution for comparison.

36. *Radioactive counting experiment. Application of the Poisson distribution II.* Consider an experiment in which the number of alpha particles emitted per second by 2 g of a radioactive substance are counted. Given that the average number of alpha particle counts per second is 5.7, calculate the probability $P(\leq 2)$ that no more than 2 alpha particles are counted in a 1 second interval.

37. *Calculation of probabilities of a Gaussian distributed random variable.* Let X be a random variable described by a Gaussian probability density with mean $\mu = 5$ and standard deviation $\sigma = 3$. Calculate the probability $P(2 \leq X \leq 11)$ that X has a value in the interval $2 \leq X \leq 11$ using (a) Eq. (5.15) with $p(x)$ given by Eq. (5.72), and (b) Eq. (5.80).

38. *Gaussian approximation to the binomial distribution.* A coin is tossed 20 times. The random variable X = *number of heads*. Calculate the probability $P(14 \leq X \leq 17)$ that either 14, 15, 16 or 17 heads are thrown using (a) a binomial distribution, and (b) a Gaussian approximation.

———————————————— **END** ————————————————

Chapter 6
Use of Computers

In this chapter we want to show how to use computer software packages to calculate quantities from measured values in experiments, to calculate standard errors in the mean, to represent data graphically and to draw graphs of the best straight line. Experimental results are most usefully represented by linear graphs. For this, we will use the method of least squares to find the best straight line and associated errors. To represent distributions of frequencies and relative frequencies of measured values, bar charts or histograms are very useful, so we will show how to produce these. In addition we will show how to fit curves to these distributions.

Specifically we will consider four software packages: ©*Microsoft Excel*, ©*Maple*, ©*Mathematica* and ©*Matlab*.[1] Of course, many other software packages are available, with a number specific to plotting graphs. Once this chapter is mastered, it should not be difficult to transfer the methods shown here to other packages. We have chosen *Excel*, a spreadsheet package, because it is commonly available and its table orientated format makes calculations very easy. We have chosen *Maple*, *Mathematica* and *Matlab* because of their tremendous mathematical and graphical power.

Excel produces charts and line graphs fairly automatically and simply. Plotting functions in Excel is a little limited. The mathematical packages, *Maple*, *Mathematica* and *Matlab*, offer much more control and flexibility in producing charts and graphs. Some might prefer to use *Excel* for calculation and then use the mathematical packages to plot the results.

Computers are great time savers, but it is important to first master calculations with a calculator and graph plotting by hand before adopting their use. Computers alone cannot produce a well-thought out presentation of results. This can only be achieved by a thorough understanding of the mathematical concepts and a good understanding of how to construct a graph (as described in earlier chapters) or chart.

---

[1] *Excel, Maple, Mathematica* and *Matlab* are registered trademarks of The Microsoft Corporation, Waterloo Maple Inc, Wolfram Research Inc, and The MathsWorks Inc, respectively.

© Springer International Publishing AG, part of Springer Nature 2018
P. N. Kaloyerou, *Basic Concepts of Data and Error Analysis*,
https://doi.org/10.1007/978-3-319-95876-7_6

It is our view that the best way to learn how to apply software packages to the solution of problems is by example. To this end we solve two examples using each of the software packages in turn. The first example involves plotting bar charts/histograms, frequency curves and relative frequency curves to represent the frequencies and relative frequencies of measured building heights. Here we will present a number of graphical options which control the final appearance of a chart or graph. The example also involves the calculation of important statistical quantities, namely, the mean, the variance, the standard deviation and the standard error in the mean.

The second example is an experiment to determine the acceleration due to gravity by measuring the times for an object to fall through various heights. This example demonstrates how to use the method of least squares to calculate the slope of the best straight line and where it cuts the y-axis, together with their errors, and then to use the slope of the best straight line to calculate the acceleration due to gravity. The graph of the best straight together with the data points will be given. Since this is a case of a line through the origin, there are two ways to draw the best straight line. One way, and the easiest, is to draw the line with the best slope through the origin. The other is to use both the slope of the best straight line and where it cuts the y-axis to plot the line.

For each of the mathematical packages we will write a separate program for each example. For *Excel* we will solve each example as a spread sheet. We will present each program in as close a format as possible to the package's format. For *Maple* and *Mathematica* we will suppress all but essential output, including charts or graphs included only to exemplify various alternative graphical options. The charts or graphs directly relevant to the examples will be exported from the program and incorporated into the chapter text. Hence, charts and graphs may not appear in the same position as when the programs are ran within *Maple* or *Mathematica*. Both *Maple* and *Mathematica* have interactive interfaces with the output from the command line produced in the line immediately following the command unless output is suppressed. *Matlab* offers both a *Command Window* and an *Editor Window*. The *Command Window* is an interactive window similar to that of *Maple* and *Mathematica*. The *Editor Window* allows a program to be written much like a Fortran program. After running the program, the output appears in the *Command Window*, while the graphical output appears in a separate graphics window.

The mathematical packages require a knowledge of various commands and options for calculations and for producing charts and graphs. Though the commands of each of the mathematical packages have similarities there are crucial differences in syntax. We will explain each command or option within the program itself either just before use or, mostly, just after use. This should serve to explain the meaning of each command and option in the command line. The command line itself will serve to show the syntax of each command and option. Where necessary, we will add a more detailed explanation of a command or option. Where more information on a command or option is desired, or to look up more options, reference can be made to the package's *help*. *Matlab* needs a bit more explanation, so we will give an introduction to the use of *Matlab* in the chapter text before presenting the *Matlab* program. For *Excel*, we will present the spreadsheet. The results of calculation cells are produced by hidden formulae. The address of each cell is specified by a row number and a column letter, e.g., C3 specifies column C, row 3. By reference to the cells address we will show the hidden formula in the body of the chapter text.

For each example, we will comment on the results after the solutions of each of the four packages have been presented, though some comments will be included either in the program itself or in the program section of the chapter text.

For the mean, variance and standard deviation we will use formulae (5.39), (5.44) and (5.46):

$$\langle X \rangle = \frac{\sum_{i=1}^{n} f(x_i)x_i}{n} = \sum_{i=1}^{n} p(x_i)x_i$$

$$\sigma^2 = \frac{\sum_{i=1}^{n} f(x_i)(x_i - \mu)^2}{n}$$

$$\sigma = \left[\frac{\sum_{i=1}^{n} f(x_i)(x_i - \mu)^2}{n} \right]^{\frac{1}{2}}$$

The formulae we will need for the method of least-squares are (4.5), (4.6) and (4.10) to (4.13):

$$m = \frac{\sum_{i=1}^{n} (x_i - \bar{x})y_i}{\sum_{i=1}^{n} (x_i - \bar{x})^2}$$

$$c = \bar{y} - m\bar{x}$$

$$\Delta m = \left[\frac{\sum_{i=1}^{n} d_i^2}{D(n-2)} \right]^{\frac{1}{2}}$$

$$\Delta c = \left[\left(\frac{1}{n} + \frac{\bar{x}^2}{D} \right) \cdot \frac{\sum_{i=1}^{n} d_i^2}{(n-2)} \right]^{\frac{1}{2}}$$

$$D = \sum_{i=1}^{n} (x_i - \bar{x})^2$$

$$d_i = y_i - Y_i = y_i - mx_i - c$$

For lines through the origin we set $c = 0$ and use the formulae (4.14) to (4.16):

$$m = \frac{\sum_{i=1}^{n} x_i y_i}{\sum_{i=1}^{n} x_i^2}$$

$$\Delta m = \left[\frac{1}{(n-1)} \cdot \frac{\sum_{i=1}^{n} d_i^2}{\sum_{i=1}^{n} x_i^2} \right]^{\frac{1}{2}}$$

$$d_i = y_i - Y_i = y_i - mx_i$$

As we mentioned in Sect. 4.3, even when we know that the line passes through the origin it can still be useful to calculate c since the amount by which the best line misses the origin gives a visual indication of errors, particularly systematic errors. In our second example, the line passes through the origin. We will use both the full formula and the 'through the origin' formulae to calculate the best line.

All four software packages have inbuilt formulae for the mean, variance and standard deviation. Note though, that the formulae refer to the sample variance and sample standard deviation. In what follows, we prefer to enter the above formulae by hand.

Before proceeding, it is perhaps worth noting that a histogram is a chart consisting of contiguous columns with widths proportional to the class interval (continuous data is divided into intervals, such that all data in an interval is regarded as a class, and the number of data in the interval is counted as the frequency of that class) and with areas proportional to the relative frequencies of the data. A bar chart is a histogram in which the class intervals are equal for continuous data, or, in the case of discrete data, each bar represents the frequency of each value of the discrete data.

For Example 1 we will show how to fit a curve to the charts and/or data points. The fitted curve will not necessarily be that good given that we are considering only a small number of data points, but we include it by way of example. Curve fitting methods rely on selecting a suitable equation, then finding the coefficients that produce the curve that most closely fits the data. The least squares method extended to include curves as well as straight lines uses polynomials of various degrees. *Maple*, *Mathematica* and *Matlab* provide both commands and a graphical interface for curve fitting. In *Maple* we have used the graphical interface, while for *Mathematica* and *Matlab* we used their curve fitting commands: the *Fit* command for *Mathematica* and the *polyfit* command for *Matlab*. Their use is described in the respective programs.

6.1   Example 1. Building Heights

Example 1 concerns the analysis of 16 repeated measurements of the height of a building. The 16 measured heights in metres are

$$33.45, 33.46, 33.47, 33.50,$$
$$33.49, 33.51, 33.48, 33.52,$$
$$33.47, 33.48, 33.49, 33.47,$$
$$33.51, 33.50, 33.48, 33.54.$$

(i) Calculate the standard error in the mean.
(ii) Draw a bar chart or histogram and joined data point graphs of the frequencies and relative frequencies of the building heights.

(iii) Use *Maples*'s graphical curve fitting interface to produce a frequency curve fitted to a frequency histogram.
(iv) Use *Mathematics*'s *Fit* command to produce a frequency curve fitted to a frequency bar chart.
 (v) Use *Matlab*'s *Fit* command to produce a frequency curve fitted to the frequency data points.

We will solve Example 1 using each of the four software packages. Once all four solutions are obtained some comments on the results will follow.

The aim of Example 1, as indicated in the example, is to show how to use the four packages (i) to calculate the mean, variance, standard deviation and standard error in the mean, (ii) to produce bar charts and histograms, (iii) to plot data points and (iv) for curve fitting.

6.1.1 Solution of Example 1 Using Excel

The calculation of the mean height together with the standard error in the mean is done in the spreadsheet shown in Fig. 6.1. The frequency and relative frequency bar charts and line graphs are shown in Fig. 6.2.

Refer to Fig. 6.1 of the *Excel* spreadsheet solution. The first (blue) row labels the columns, while the first (blue) column labels the rows. The measured building heights are entered in column B, rows 2 to 17 or, more conveniently stated, cells B2 to B17. In cell B18 we entered a formula to calculate the mean, while cell B19 is a text cell in which we have rounded the mean by hand.

To enter a formula in a cell, activate the cell (by double clicking the left mouse button) and type an equal sign followed by the formula. The formula appears in one of the header text bars, while the result of the formula appears in the cell. The formula in cell B18 looks like this,

$$= \text{SUM}(\text{B2} : \text{B17})/16.$$

The formula adds the contents of cells B2 to (the colon acts as 'to') B17 and divides the answer by 16. Typically, and conveniently, formulae are written in terms of cells rather than numbers. Below, we give the formulae for each cell that contains a formula. With the mean calculated, the residuals can be calculated and this is done in column C, rows C2 to C17. The formula for the residual in cell C2 is

$$= \text{B2} - 33.48875,$$

where 33.48875 is the mean calculated in cell B18. Instead of laboriously typing the formula in the rest of the cells in the column C, cell C2 can be highlighted, copied and its contents pasted to the remaining cells C3 to C17. *Excel* will adjust the row

	A	B	C	D	E	F	G	H
		SOLUTION OF EXAMPLE 1. MEASURED BUILDING HEIGHTS						
1		**HEIGHT (m) ASCENDING ORDER**	**RESIDUALS**	**RESIDUALS SQUARED**		**HEIGHT (m) ASCENDING ORDER**	**FREQUENCY**	**RELATIVE FREQUENCY**
2	1	33.45	-0.0387500	0.00150156250		33.45	1	0.0625
3	2	33.46	-0.0287500	0.00082656250		33.46	1	0.0625
4	3	33.47	-0.0187500	0.00035156250		33.47	3	0.1875
5	4	33.47	-0.0187500	0.00035156250		33.48	3	0.1875
6	5	33.47	-0.0187500	0.00035156250		33.49	2	0.125
7	6	33.48	-0.0087500	0.00007656250		33.5	2	0.125
8	7	33.48	-0.0087500	0.00007656250		33.51	2	0.125
9	8	33.48	-0.0087500	0.00007656250		33.52	1	0.0625
10	9	33.49	0.0012500	0.00000156250		33.54	1	0.0625
11	10	33.49	0.0012500	0.00000156250			16	1
12	11	33.5	0.0112500	0.00012656250				
13	12	33.5	0.0112500	0.00012656250				
14	13	33.51	0.0212500	0.00045156250				
15	14	33.51	0.0212500	0.00045156250				
16	15	33.52	0.0312500	0.00097656250				
17	16	33.54	0.0512500	0.00262656250				
18		33.488750		0.00052343750				
19		**MEAN=33.49**		**VARIANCE= 0.00052**				
20				STANDARD DEVIATION			0.0228787565	
21				STANDARD ERROR IN THE MEAN			0.00590727	
22				STANDARD ERROR IN THE MEAN (2 dc pl)			0.01	
23								
25		Answer. The height of the building = 33.49 ± 0.01 m						

Fig. 6.1 *Excel* spreadsheet to calculate the height of a building and the standard error in the mean

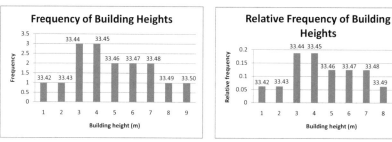

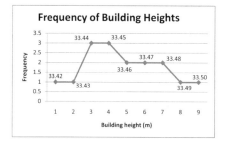

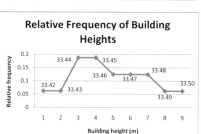

Fig. 6.2 *Excel* bar charts of the frequencies and relative frequencies of the building heights, together with joined data point plots of the frequencies and relative frequencies of the building heights

numbers accordingly. The formula for the square of the residual in cell D2 has the form

$$= C2 * C2,$$

and is repeated in cells D3 to D17 with cell addresses adjusted accordingly. The variance is given in cell D18:

$$= SUM(D2 : D17)/16$$

Cell F20 gives the standard deviation:

$$= SQRT(D18)$$

Cell F21 gives the required standard error in the mean:

$$= F20/SQRT(16 - 1)$$

Cell F22 rounds off the standard error in the mean to 2 dc. pl.

$$= ROUND(F21, 2)$$

The various plots are made from column G, rows 2 to 10 containing the frequencies, and from column H, rows 2 to 10, containing the relative frequencies. Column H is obtained by dividing column G by the total number of measurements (16) as, for example, in cell H2:

$$= G2/16$$

In cell G11 the frequencies are summed to check that they sum to the total number of measurements (16):

$$= SUM(G2 : G10)$$

In cell H11 the relative frequencies are summed to check they sum to 1:

$$= SUM(H2 : H10)$$

From Fig. 6.1 we get the mean building height and its standard error in the mean:

Answer. The height of the building $= 33.49 \pm 0.01$ m.

The original measurements were made to 4 significant figures (sf) so that the calculated mean cannot be more accurate than this. Hence, the standard error in the mean must be rounded to 2 decimal places so that it corresponds to the 4th sf of the mean height. The answer is therefore given to 4 sf with the error of the 4th sf indicated.

The *Excel* inbuilt functions AVERAGE and STDV can be used to find the mean and sample standard deviations, respectively. We have preferred to enter the formula by hand, first, because it is more instructive, and second, because it is the standard deviation that is needed to calculate the standard error in the mean, not the sample standard deviation.

To produce a chart, follow these steps:

1. Highlight the column or row containing the data to be plotted. If more than one row or column is highlighted each set of data is plotted on the same axes.
2. Select (by left clicking the mouse button) the top menu bar option *Insert*. A sub-menu bar appears containing the group *Charts*. The *Charts* group offers the following choice of charts: *Column, Line, Pie, Bar, Area, Scatter* and *Other Charts*. Note that reference is not made to histograms. What are called *Column* charts are bar charts, while what are called 'bar charts' are horizontal bar charts.
3. Select the type of chart desired. For example, select *Column*. A further menu appears offering choices such as *2-D Column, 3-D Column* etc.
4. Select *2-D Column*. The chart is automatically produced.
5. Once the chart is produced, a number of format options appear in the top menu bar. These include: colour of the columns, title bar, axes labels, a legend, grid, and labelling of the columns. A variety of positions and combinations are offered.

The frequency charts of Fig. 6.2 were produced by highlighting (selecting) cells G2 to G10 containing the frequencies and selecting *Column* for the first frequency chart and *Line* for the second frequency chart. The relative frequency chart and line chart of Fig. 6.2 are similarly produced by highlighting cells H2 to H10 containing the relative frequencies.

6.1.2 Solution of Example 1 Using Maple

The solution is presented as a *Maple* program. Explanations are included within the program. The meaning of *Maple* commands are more-or-less obvious from their usage. Therefore, our approach will be to state what the command line is supposed to do. This should serve to explain the meaning of the command. We will add further explanation if and where needed. The usage and meaning of a number of graphics options is less obvious, so we have included more detailed explanations of these.

Alternatively to controlling the appearance of a chart or graph by command line options as above, *Maple* offers a graphics menu bar which allows the options to be implemented by choosing the appropriate options. The graphics menu bar is invoked by right clicking the mouse button on the graphic. (We will repeat these instructions within the *Maple* program that follows.)

The following is the *Maple* program. In *Maple*, text lines begin with '[', while command input lines begin with '[>'. The output of a command is in italic blue type. Output lines are numbered. We will suppress the output of most commands, including those for charts or graphs included only to illustrate the use of graphical options. Only command line output relevant to the solution of Example 1 will be shown.

[*MAPLE* PROGRAM FOR THE SOLUTION OF EXAMPLE 1. MEASURED BUILDING HEIGHTS

[To plot bar charts and histograms, the *plots* package needs to be loaded, for statistics, the *Statistics* package needs to be loaded, and for curve fitting, the *CurveFitting* package needs to be loaded.

[> *restart*

[The *restart* command clears memory. It is always a good idea to start a program with this command.

[> *with*(*plots*):
[> *with*(*Statistics*):
[> *with*(*CurveFitting*):

[The data points, the measured building heights, are entered as a list labelled *bheightsi*:

[> *bheightsi* :=[33.45, 33.46, 33.47, 33.50, 33.49, 33.51, 33.48, 33.52, 33.47, 33.48, 33.49,33.47, 33.51, 33.50, 33.48, 33.54]:

[The right-hand-side is a Maple list.

[Note that terminating a command line with a colon suppresses output, while leaving a command line open or terminating with a semicolon shows output.

[

[The sort command is used to place the measurements in ascending order

[> *bheights* := *sort*(*bheightsi*)
  bheights :=[33.45, 33.46, 33.47, 33.47, 33.47, 33.48, 33.48, 33.48, 33.49, 33.49, 33.50, 33.50, 33.51, 33.51, 33.52, 33.54]   (1)

[

[The sum command is used to sum the measurements

[> *bheightssum* := *sum*(*bheights*[*i*], *i* = 1..16)
$$bheightssum := 535.82 \qquad (2)$$

[*bheights*[*i*] refers to the *i*th item of the list, e.g.,

[> *bheights*[3]
$$33.47 \qquad (3)$$

[The two dots '..' stand for 'to', e.g., 1..16 means 1 to 16

[

[Dividing by the total number 16 of measurements gives the mean

[> $bheightsmean := \dfrac{bheightssum}{16}$
$$bheightsmean := 33.48875000 \qquad (4)$$

[

[Subtracting each measurement from the mean gives the residuals

[> *bheightsresidulas* := [33.45 − 33.48875, 33.46 − 33.48875, 33.47 − 33.48875, 33.47 − 33.48875, 33.47 − 33.48875, 33.48 − 33.48875, 33.48 − 33.48875, 33.48 − 33.48875, 33.49 − 33.48875, 33.49 − 33.48875, 33.50 − 33.48875, 33.50 − 33.48875, 33.51 − 33.48875, 33.51 − 33.48875, 33.52 − 33.48875, 33.54 − 33.48875]:

[

[The following is a check that *bheightsresiduals* sums to zero

[> *bheightsresidualsum* := *sum*(*bheightsresiduals*[*k*], *k* = 1..16)
$$bheightsresidualssum := 0 \qquad (5)$$

[

[Taking the squares of the residuals gives

[> *bheightsresidualssq* := [(33.45 − 33.48875)², (33.46 − 33.48875)², (33.47 − 33.48875)², (33.47 − 33.48875)², (33.47 − 33.48875)², (33.48 − 33.48875)²,

$(33.48 - 33.48875)^2$, $(33.48 - 33.48875)^2$, $(33.49 - 33.48875)^2$, $(33.49 - 33.48875)^2$,
$(33.50 - 33.48875)^2$, $(33.50 - 33.48875)^2$, $(33.51 - 33.48875)^2$, $(33.51 - 33.48875)^2$,
$(33.52 - 33.48875)^2$, $(33.54 - 33.48875)^2$]:

[

[Determination of the sum of the residuals squared:

[> $bheightsresidualssqsum := sum(bheightsresidualssq[m], = m = 1..16)$

$$bheightsresidualssqsum := 0.008375 \qquad (6)$$

[

[Taking the mean of $bheightsresidualsqsum$ gives the variance

[> $bheightsvar := \dfrac{bheightsresidualssqsum}{16}$

$$bheightsvar := 0.0005234375 \qquad (7)$$

[

[The square root of the variance gives the standard deviation

[> $bheightsstdv := \mathrm{sqrt}(bheightssvar)$

$$bheightsstdv := 0.02287875652 \qquad (8)$$

[

[Determination of the standard error in the mean:

[> $bheightserror := \dfrac{bheightsstdv}{(16-1)^{0.5}}$

$$bheightserror := 0.005907269533 \qquad (9)$$

[

[The standard error in the mean rounded to 1 sf is

[> $evalf(\%, 1)$

$$0.006 \qquad (10)$$

[
[**The standard error in the mean to 2dc. pl. is 0.01**
[**Answer. Height of the building = 33.49 + / − 0.1 m**
[The original measurements were made to 4 significant figures (sf) so that the calculated mean cannot be more accurate than this. Hence, the standard error in the mean
must be rounded to 2 decimal places so that it corresponds to the 4th sf of the mean
height. The answer is therefore given to 4 sf with the error of the 4th sf indicated.

[

[Maple has inbuilt functions for the mean, variance and standard deviation, namely,
Mean(), *Variance*() and *StandardDeviation*(), respectively. But, it should be noted that
the latter two functions deliver the sample variance and sample standard deviation. We
have preferred to enter the formula by hand, first, because it is more instructive, and
second, because it is the standard deviation that is needed to calculate the standard
error in the mean, not the sample standard deviation.

[

[# The Charts

[There are a variety of options for controlling the appearance of charts or graphs. We
will use the various options first and then explain each one after. With some exceptions,
options apply to all types of charts and graphs. The options we present are, of course,
by no means comprehensive.

[
[# Histograms
[Histograms are produced with the Histogram command which plots the frequencies of a list of data. In our case the list is *bheights*. The frequency of data is automatically counted to produce the histogram . Because of this automation, the Histogram will not be used to represent relative frequencies. To plot the relative frequencies *pointplot* (see below) will be used.
[

[The histogram of the frequencies in the list *bheights* is obtained by setting $frequencyscale = absolute$ (explained below):

[> $PP1 := Histogram(bheights, frequencyscale = absolute,$
$view[33.43..33.54, 0..4], axesfont = [Calibri, roman, 12],$
$titlefont = [Calibri, roman, 18], labelfont = [Calibri, roman, 14],$
$title =$ "Frequency of Building Heights", $axes = frame,$
$labels = [$"Building heights (m)", "Frequency"$],$
$labeldirections = [horizontal, vertical], color =$ "Orange",
$axis[1] = [thickness = 1.5], axis[2] = [thickness = 1.5]) :$

[> $display(PP1)$
]

[Graph $PP1$ is shown in Fig. 6.3. When the program is ran, graphs are positioned immediately following the command that produces them. Here, the positions of the graphs will not, in general, follow the commands that produced them.

[
[# Some Options
[*frequencyscale* = t, where t = *absolute* (to plot the frequency) or *relative* (to plot the relative frequency). We did not get good results using the choice $frequencyscale = relative$, so we will plot the relative frequencies using *pointplot*.

$view = [x_{min}..x_{max}, y_{min}..y_{max}]$ - specifies the $x-$ and $y-$ axis range. The two dots '..' stand for 'to', e.g., $x_{min} .. x_{max}$ is read x_{min} to x_{max}.

[$font = [font name, face, point size]$ - face can be *roman*, *italic* or *bold* among others, while point size can be, for example, 10pt, 12pt etc.. The *font* command specifies the font for elements of the plot.

[$titlefont = [font name, face, point size]$ - specifies the font to be used for the plot title. It overrides the specifications in *font*.

[$labelfont = [font name, face, point size]$ - specifies the font to be used for the axes labels. It overrides the specifications in *font*.

[$axesfont = [font name, face, point size]$ - specifies the font to be used for the axes numbers. It overrides the specifications in *font*.

[*title* = "Title" - Notice that the title should be enclosed with speech marks.

[*axes* = t , where t = *boxed*, *frame*, *none* or *normal*.

[*axis* = t - axis option t is applied to all axes. The t-option of interest here is t = *thickness*, or, if grid lines are desired t = *gridlines* = [*options*], where some options are [*number of grid lines*, *linestyle* = *dot*, *color* = *blue*, *thickness* = 2]. Note also that *gridlines* = *default* attributes to both axes the default values.

[The horizontal and vertical grid lines can be given individual options using *axis*[*dir*] = *gridlines* =[number of grid lines, *color* = colour name], *thickness* = number between 0 and 1], where *dir* = 1 = horizontal axis, while *dir* = 2 = vertical axis. An example is:

[> *PP2* := *Histogram*(*bheights*, *frequencyscale* = *absolute*,
   view[33.43..33.54, 0..4], *axesfont* = [*Calibri*, *roman*, 12],
   titlefont = [*Calibri*, *roman*, 18], *labelfont* = [*Calibri*, *roman*, 14],
   title = "Frequency of Building Heights", *axes* = *frame*,
   labels = ["Building height (m)", "Frequency"],
   labeldirections = [*horizontal*, *vertical*], *color* = "Orange",
   axis[1] = [*gridlines* = [10, *color* = *red*], *thickness* = 0],
   axis[2] = [*gridlines* = [10, *color* = *blue*], *thickness* = 0]) :

[> *display*(*PP2*)
   ]

[Graph *PP2* is shown in Fig. 6.4.

[*thickness* =t , where t is positive integer. t=0 represents the thinnest line.

[*labels* = ["*x*-axis label", "*y*-axis label"] - again, notice the speech marks, they indicate a character string.

[*labeldirections* = [*horizontal*, *vertical*] - specifies that the *x*-axis label should be horizontal, while the *y*-axis label should be vertical.

[*style* = *polygon* - gives a different appearance to the bars.

[*color* = "name of the colour" - specifies the colour of the bars. Notice that the colour names must be enclosed in speech marks. Examples of colour names are "Red", "Blue", "Orange" among many others. Notice also that the colour names must be capitalised.

[

[ALTERNATIVELY TO CONTROLLING THE APPEARANCE OF A CHART OR GRAPH BY COMMAND LINE OPTIONS AS ABOVE, *MAPLE* OFFERS A GRAPHICS MENU BAR. THE GRAPHICS MENU BAR IS INVOKED BY RIGHT CLICKING THE MOUSE BUTTON ON THE GRAPHIC.

[Line Plots of Frequency and Relative Frequency Curves

[Data points can be plotted using *pointplot* by defining a list of coordinates to be plotted.

[To plot a frequency curve we take the heights as the *x*-coordinates, with the frequencies as the *y*-coordinates. We write each coordinate as a list, e.g., [33.47,3], then make a list of all the coordinate lists, i.e., we nest the coordinate lists in a larger list, which we label *bheightspnt*:

[> *bheightspnt* :=[[33.45, 1], [33.46. 1], [33.47, 3], [33.48, 3], [33.49, 2], [33.50, 2], [33.51,2], [33.52, 1], [33.54, 1]]:

[

[Data plot, with data points joined by a line, of the frequencies of the heights in `bheights`

[> $PP3 := pointplot(bheightspnt, \ style = pointline,$
$symbol = circle, \ view[33.43..33.54, 0..4], \ axesfont = [Calibri,$
$roman, 12],$
$titlefont = [Calibri, roman, 18], \ labelfont = [Calibri, roman, 14],$
$title =$ "Frequency of Building Heights", $axes = frame,$
$labels = [$"Building heights (m)", "Frequency"$],$
$labeldirections = [horizontal, vertical], color =$ "Orange") :

[> $display(PP3)$

[Graph $PP3$ is shown in Fig. 6.5.

[Dividing the frequencies in $bheightspnt$ by 16, defines a new coordinate list $bheightspntrf$

[> $bheightspntrf := [[33.45, \frac{1}{16}], [33.46, \frac{1}{16}], [33.47, \frac{3}{16}], [33.48, \frac{3}{16}],$
$[33.49, \frac{2}{16}],$
$[33.50, \frac{2}{16}], [33.51, \frac{2}{16}], [33.52, \frac{1}{16}], [33.54, \frac{1}{16}]]$:
]

[Data plot, with data points joined by a line, of the relative frequencies of the heights in `bheights`

[> $PP4 := pointplot(bheightspntrf, \ style = pointline,$
$symbol = circle, \ view[33.43..33.54, 0..3], \ axesfont = [Calibri,$
$roman, 12],$
$titlefont = [Calibri, roman, 14], \ labelfont = [Calibri, roman, 14],$
$title =$ "Relative Frequency of Building Heights", $axes = frame,$
$labels = [$"Building heights (m)", "Relativefrequency"$],$
$labeldirections = [horizontal, vertical], color =$ "Orange") :
[

[Graph $PP4$ is shown in Fig. 6.6.

[ ]

Some Additional Options for *pointplot*

[All the options defined above can be used with *pointplot*. Here are some additional ones:

[*style*=s, where s = *line*, *point*, or *pointline* are the options of interest to us here.

[*symbol* = s - specifies the symbol for data points. The possible choices are s= *asterix*, *box*, *circle*, *cross*, *diagonalcross*, *diamond*, *point*, *solidbox*, *solid*, *circle* or *soliddiamond*

[
[# Curve Fitting
[By way of example, we will try to fit a curve to the frequency data points listed in *bheightspnt*.
[We use *Maple*'s Interactive command which invokes a graphical interface to enable curve fitting. Information on how to use the graphical curve fitting interface can be found under curve fitting in *Maple*'s excellent *Help*. There are two ways of using the *Interactive* command. First, the data points are included in the command. Second, the data points are left out, in which case the first graphic that appears is a table to enter the data points. Here, we will invoke the graphical interface using the *interactive* command with the data points included . Better results are obtained if values of the heights are replaced by their position in the list *bheights*, with repeated values counted as being in the same position. Thus, the list of coordinates *bheightspnt*, [[33.45,1], [33.46,1], [33.47,3], [33.48,3], [33.49,2], [33.50,2], [33.51,2], [33.52, 1] , [33.54,1]] is mapped onto the following coordinates [[1,1] , [2,1], [3,3], [4,3], [5,2], [6,2], [7,2], [8, 1] , [9,1]]. A number of curve fitting options are offered. We found the *least squares* option with the polynomial $ax^8 + bx^7 + cx^3 + d^2$ gave the best curve. There is a choice of returning a graph or the coefficients of the polynomial. We preferred to have the values of the 'best fit' coefficients returned.
[The following command invokes the graphical interface
[> *Interactive* ([[1,1], [2,1], [3,3], [4,3], [5,2], [6,2], [7,2], [8,1], [9,1], x)
[By selecting the *interpolant* of the *on 'Done' return* option, the plot interpolant function is placed in the worksheet.
[Definition of the interpolant function using the coefficients of the interpolant returned by *Interactive*:
[> $INTERPBH := (x) \rightarrow \frac{2452318603875385298701}{4381113840796886320764}x^2 - \frac{22961214042980559683}{231804965121528376760}x^3$

$+ \frac{1615232083666614959}{6954148953645851 30280}x^7 + \frac{4238492937391799}{219055692039844 3160382}x^8$
[Note the syntax for defining a function in *Maple*.
[Plot of fitted curve
[> $PP5 := plot(INTERPBH(x), x = 0..10, view[0..10, 0..3.5],$
$axesfont = [Calibri, roman, 12], titlefont = [Calibri, roman, 18],$
$labelfont = [Calibri, roman, 14],$
$title = $ "Frequency of Building Heights", $axes = frame,$
$labels = [$"Building heights (m)", "Frequency"],
$labeldirections = [horizontal, vertical], color = $ "blue",
$axis[1] = [thickness = 1.5], axis[2] = [thickness = 1.5]):$
[We have suppressed the display of the plot as it is more interesting to combine the fitted curve *PP5* with the frequency histogram. To do this the $x-$ and $y-$axis scales should match. This can be achieved by first creating a new list, *bheieghtsp*, by replacing each height by its position

Fig. 6.3 *Maple* histogram of the frequencies of the building heights

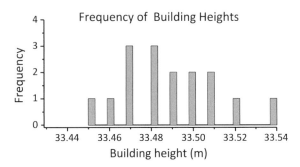

Fig. 6.4 An example to include grid lines in a *Maple* chart (or graph). In this example, horizontal and vertical grid lines are given individual options

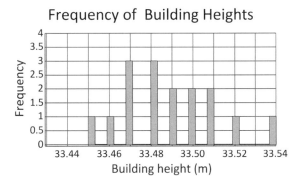

in the list *bheights* and writing its position a number of times equal to its frequency, then plot a histogram of *bheightsp*

[> *bheights* := [1, 2, 3, 3, 3, 4, 4, 4, 5, 5, 6, 6, 7, 7, 8, 9] :

[

[Plot of a histogram of the frequencies of the heights in *bheightsp*

[> *PP6* := *Histogram*(*bheightsp*, *frequencyscale* = *absolute*,
   view[0..10, 0..3.5], *axesfont* = [*Calibri*, *roman*, 12],
   titlefont = [*Calibri*, *roman*, 18], *labelfont* = [*Calibri*, *roman*, 14],
   title = "Frequency of Building Heights", *axes* = *frame*,
   labels = ["Heights position label", "Frequency"],
   labeldirections = [*horizontal*, *vertical*], *color* = "Orange",
   axis[1] = [*thickness* = 1.5], *axis*[2] = [*thickness* = 1.5]):

[Again we have not displayed graphic *PP6*, but will instead display the combined plot of *PP5* and *PP6*:

[> *display*(*PP5*, *PP6*)

[

[The combined *PP5* and *PP6* plot is shown in Fig. 6.7.

[As can be seen, the curve fitting is reasonably good, though the end part of the curve deviates from the Gaussian shape expected for measurements of this type. This is due to the small number of measurements made.

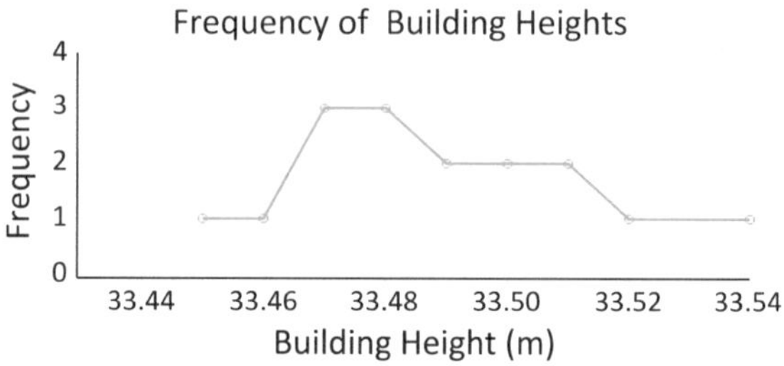

Fig. 6.5 *Maple* data plot, with data points joined by a line, of the frequencies of the building heights

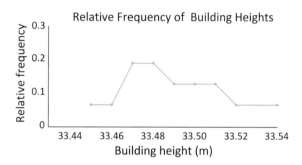

Fig. 6.6 *Maple* line and data plot of the relative frequencies of the building heights

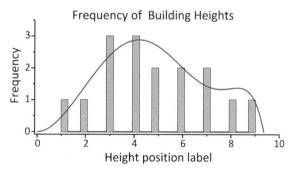

Fig. 6.7 *Maple* histogram of the frequencies of the building heights together with the fitted frequency curve

6.1.3   Solution of Example 1 Using Mathematica

The solution is presented as a *Mathematica* program. Explanations are included within the program. The meaning of Mathematica commands are more-or-less obvious from their usage. Therefore, our approach, as with Maple, will be to state what the command line is supposed to do. This should serve to explain the meaning of the command. We will add further explanation if and where needed. The usage and meaning of a number of graphics options is less obvious, so we have included a more detailed explanation of these.

Some aspects of the appearance of charts and graphs can also be controlled by a graphics menu box named *Drawing Tools*. But, *Drawing Tools* is most useful for adding text or drawing additional graphics objects on the original plot. *Drawing Tools* is invoked as follows: Right click the mouse button on the chart or plot. This brings up an options box. Selecting (by left clicking the mouse button) the option *Drawing Tools* brings the *Drawing Tools* options box, which offers some options that control the appearance of the chart or plot. (We will repeat these instructions within the *Mathematica* program that follows.) In *Mathematica*, text lines are in regular type while input lines use a typewriter font. The output of a command is in grey type. After the program is run, input and output lines are numbered and begin with 'In[i]:=' and 'Out[i]:=', respectively. Text lines are not numbered. *Mathematica* indicates cells and cell groups by nested square brackets placed at the right of the program page. Though, in the presentation of the program that follows, we have tried to follow the format of the program within *Mathematica*, we will not include the nested square brackets. The output of a command is in italic blue type. We will suppress the output of most commands, including those for charts or graphs, included only to illustrate the use of graphical options. Only command line output relevant to the solution of Example 1, will be shown.

MATHEMATICA PROGRAM FOR THE SOLUTION OF EXAMPLE 1. MEASURED BUILDING HEIGHTS

In[1]:= `clear;`
The data points, the measured building heights, are entered as a list labelled *bheights*

In[2]:= `bheightsi ={33.45, 33.46, 33.47, 33.50, 33.49, 33.51, 33.48, 33.52, 33.47, 33.48, 33.49, 33.47, 33.51, 33.50, 33.48, 33.54};`
Note that terminating a command line with a semicolon suppresses output, while leaving a command line open shows output.
The *Sort* command is used to place the measurements in ascending order

In[3]:= `bheights = Sort[bheightsi]`
Out[3]:= `{33.45, 33.46, 33.47, 33.47, 33.47, 33.48, 33.48, 33.48, 33.49, 33.49, 33.5, 33.5, 33.51, 33.51, 33.52, 33.54}`
The *Sum* command is used to sum the measurements

In[4]:= `bheightssum = Sum[bheightsi[[i]], {i, 1, 16}] ;`
The syntax for the Sum command is : $Sum[i, \{i, i_{min}, i_{max}\}]$. *bheights*[[i]] refers to the ith item of the list, e.g.,

In[5]:= `bheights[[3]]`
Out[5]:= `33.47`

Dividing by the total number 16 of measurements gives the mean

In[6]:= `bheightsmean = bheightssum /16;`

Each measurement is subtracted from the mean to get the residuals

In[7]:= `bheightsresiduals = bheights - bheightsmean;`

The following is a check that *bheightsresiduals* sums to zero

In[8]:= `bheightsresidualssum =Sum [bheightsresiduals`
`[[j]], {j, 1, 16}] ;`

The squares of the residuals is taken and placed in the list *bheightsresidualssq*

In[9]:= `bheightsresidualssq = bheightsresiduals^2;`

Determination of the sum of the residuals squared:

In[10]:= `bheightsresidualssum = Sum[bheightsresidualssq`
`[[j]], {j, 1, 16}] ;`

The mean of *bheightsresidualssum* gives the variance

In[11]:= `bheightsvar = bheightsresidualssum/16`

Out[11]:= `0.000523437`

The square root of the variance gives the standard deviation

In[12]:= `bheightssddv = Sqrt[bheightsvar]`

Out[12]:= `0.0228788`

Determination of the standard error in the mean:

In[13]:= `bheightssterror = bheightssddv/Sqrt[16 - 1]`

Out[13]:= `0.00590727`

The standard error in the mean is rounded to 1 sf using *NumberForm*[*number*, *number of sf*], which is a *Mathematica* command for rounding to a given number of significant figures (sf).

In[14]:= `NumberForm[bheightssterror, 1]`

Out[14] //NumberForm=

`0.006`

The standard error in the mean to 2 dc. pl. is 0.01
Answer. Height of the building = 33.49 ±0.01 m

The original measurements were made to 4 significant figures (sf) so that the calculated mean cannot be more accurate than this. Hence, the standard error in the mean must be rounded to 2 decimal places so that it corresponds to the 4th sf of the mean height. The answer is therefore given to 4 sf with the error of the 4th sf indicated.

Mathematica has inbuilt functions for the mean, variance and standard deviation, namely, *Mean*[], *Variance*[] and *StandardDeviation*[], respectively. But, it should be noted that the latter two functions deliver the sample variance and sample standard deviation. We have preferred to enter the formulae by hand, first, because it is more instructive, and second, because it is the standard deviation that is needed to calculate the standard error in the mean, not the sample standard deviation.

The Charts

There are a variety of options for controlling the appearance of charts or graphics. In *Mathematica* the meaning and use of graphics options is obvious from their use in the command line. Similarly, the syntax of the options is fairly clear from the command line. More values of the options used here, and indeed, many more options, can be found in *Mathematica*'s excellent *Help*. Hence, we will only add a few extra explanatory comments here and there. With some exceptions, the options apply to all types of charts and graphs. The options we present are, of course, by no means comprehensive. Note that *Mathematica* has separate commands for histograms and bar charts. Although height is a continuous quantity, the measured values are necessarily discrete. Bar charts are preferred in the case of discrete values because the column labels refer to the discrete value and the frequency also refers to that value. Histograms tend to count data points in unequal intervals, Another advantage of bar charts is that the plots of relative frequencies can be produced in a straight forward way. We will. however, include histograms by way of example.

Histograms

Histograms are produced with the *Histogram* command which plots the frequencies of a list of data. In our case the list is *bheights*. The frequency of data is automatically counted to produce the histogram.

Histogram of frequencies of the heights in *bheights*

```
In[15]:= QQ1=Histogram[bheights, 10, ChartLabels →
        {33.45, 33.46,
        33.47, 3.48, 33.49, 33.5, 33.51, 33.52,
        33.54},
        AxesLabel → {HoldForm["Height (m)"],
        HoldForm["Frequency"]},
        PlotLabel → HoldForm["Frequency of Building
        Heights"],
        LabelStyle → {FontFamily → "Calibri", 14,
        GrayLevel[0]},
        ChartStyle → "Pastel"];
```

In[16]:= Show[QQ1]
]

Histogram *QQ*1 is shown in Fig. 6.8.

The number 10 specifies the number of bars. Obviously, this number must be chosen as an integer. *HoldForm*[*"text or expression"*] prints text (indicated by enclosing the text in speech marks) as is, and prints expressions (indicated by leaving out the speech marks) unevaluated. *Chartstyle* selects colours or colour schemes for the columns in a bar chart or histogram. *ChartStyle* → {*Red*, *Green*, *Orange*, ...} selects a color for each column. The 14pt *LabelStyle* is the point size for the axes labels and axes numbering. *BaseStyle* → {FontFamily → "Calibri", FontSize → 16} controls all elements of the plot. The point size for axes labels and plot labels can also be chosen from the *Format* menu offered in the top menu bar of the *Mathematica* window. The point size chosen from the menu bar overrides the value given in *LabelStyle*. For the histogram in Fig. 6.8, 18 pt was chosen for the plot label from the *Format* menu. *GrayLevel*[d] specifies how dark or light the objects it refers to are: d = 0 to 1, 0 = black, 1 = white. In our case, *GreyLevel* specifies that text should be black.

MATHEMATICA OFFERS A GRAPHICS MENU PALLET NAMED *Drawing Tools*. AS THE NAME IMPLIES, IT OFFERS NUMEROUS DRAWING TOOLS, SUCH AS ADDING ARROWS, TEXT, RECTANGLES ETC. *Drawing Tools* IS INVOKED AS FOLLOWS: RIGHT CLICK THE MOUSE BUTTON ON THE CHART OR PLOT. THIS BRINGS UP AN OPTIONS BOX. SELECTING (BY LEFT CLICKING THE MOUSE BUTTON) THE OPTION *DrawingTools* BRINGS UP THE *Drawing Tools* PALLET.

Bar Charts

Bar chart of the frequencies of the heights in *bheights*

In[17]:= QQ2 := BarChart[1, 1, 3, 3, 2, 2, 2, 1, 1, ChartLabels → 33.45, 33.46, 33.47, 3.48, 33.49, 33.5, 33.51, 33.52, 33.54, ChartElementFunction → "GlassRectangle", ChartStyle → "Pastel", AxesLabel → {HoldForm["Height (m)"], HoldForm ["Frequency"]}, PlotLabel → HoldForm["Frequency of Building Heights"], LabelStyle → {FontFamily → "Calibri", 12, GrayLevel [0]}, AxesStyle → Thick, TicksStyle → Directive[Thin]]
]

In[18]:= Show[QQ2]

Bar chart *QQ2* is shown in Fig. 6.9.

ChartElementFunction produces charts with different effects. This option can also be used with Histograms.

Sometimes a SUGGESTION BAR appears automatically after executing a cell. If not, it can be invoked by right-clicking the mouse on the circular arrow button that appears in the bottom right of the output after the cell is executed.

To plot the relative frequencies we first create a list of relative frequencies labelled by *bheightsrelfreq*

In[19]:= `bheigtsfreq = {1, 1, 3, 3, 2, 2, 2, 1, 1};`

In[20]:= `bheightsrelfreq = bheigtsfreq/16;`

Bar chart of relative frequencies of the heights in *bheights*.

In[21]:= QQ3=BarChart[bheightsrelfreq, ChartLabels → {33.45, 33.46, 33.47, 3.48, 33.49, 33.5, 33.51, 33.52, 33.54}, ChartElement-Function → "GlassRectangle", ChartStyle → "Pastel", AxesLabel →{HoldForm["Height (m)"], HoldForm["Relative frequency"]}, PlotLabel → HoldForm["Relative Frequency of Building Heights"], LabelStyle → {FontFamily → "Calibri", 12, GrayLevel[0]}, AxesStyle → Thick, TicksStyle → Directive[Thin]]

In[22]:= Show[QQ3]
]

Bar Chart of *QQ*3 is shown in Fig. 6.10.

Line Plots of Frequencies and Relative frequencies

The *ListLinePlot* plots lists of coordinates. The pairs of coordinates are placed in a two entry list. These lists are nested in yet another list. Most of the options used for histograms and bar charts can also be used for line plots. The additional option *Mesh → All* adds data points to the line plot.

Line and data plot of frequencies of the heights in *bheights*

In[23]:= QQ4=ListLinePlot[{ {33.45, 1}, {33.46, 1}, {33.47, 3}, {33.48, 3}, {33.49, 2}, {33.5, 2}, {33.51, 2}, {33.52, 1}, {33.54, 1} }, Mesh → All, PlotStyle → {Orange, PointSize[Large]}, AxesLabel → {HoldForm["Height (m)"], HoldForm["Frequency"]}, PlotLabel → HoldForm["Frequency of Buiding Heights"], LabelStyle → {FontFamily → "Calibri", 12, GrayLevel[0]}, AxesStyle → Thick, TicksStyle → Directive[Thin]]

In[24]:= Show[QQ4]
]

Graph *QQ*4 is shown in Fig. 6.11.

PlotStyle → {*style*1, *style*2, ...} specifies how plot objects such as lines and points are drawn. The style can be the colour, line thickness, point size etc.

PointSize[*d*] specifies the point sizes. *d* can be a number, but conveniently *d* = *Tiny*, *Small*, *Medium*, or *Large* are also options.

Grid lines can be included using the *GridLines* command. The simplest way is the choice *Gridlines* → *Automatic*. The colour and style (e.g., dashed or dotted among other styles) of the grid lines can be chosen using the command *GridLinesStyle* as the following plot shows:

Line and data plot of the relative frequencies of the heights in *bheights*

In[25]:= QQ5=ListLinePlot[{ {33.45, 1/16}, {33.46, 1/16}, {33.47, 3/16}, {33.48, 3/16}, {33.49, 2/16}, {33.5, 2/16}, {33.51, 2/16}, {33.52, 1/16}, {33.54, 1/16} },

Mesh → All, PlotStyle → {Blue, PointSize[Large]},

AxesLabel → {HoldForm["Height (m)], HoldForm["Relative frequency"]},

PlotLabel → HoldForm["Relative Frequency of Building Heights],

LabelStyle → {FontFamily → "Calibri", 12, GrayLevel[0]},

AxesStyle → Thick, TicksStyle → Directive[Thin]]

In[26]:= Show[QQ5]
]

Graph *QQ*5 is shown in Fig. 6.12.

Curve Fitting

The *Mathematica* curve fitting command is *Fit*, with syntax:

Fit[*data*, *fit function*, *fit function variable*], where *data* = the data to fit, *fit function* = polynomial chosen for the fit, and *fit function variable* = the variable in which the polynomial is expressed.

The *Fit* command is based on the least-squares method which uses polynomials as the fit-functions. For example, a linear fit uses $ax + b$, a quadratic fit uses $ax^2 + bx + c$, while a cubic fit uses $ax^3 + bx^2 + cx + d$. The fit polynomial is entered as either linear = {1,x}, quadratic={$1, x, x^2$}, cubic={$1, x, x^2, x^3$} etc. as shown in the following *Fit* command, which fits a curve to the frequency data points in the list *bheigtsfreq*.

In[27]:= bestcurve = Fit[bheigtsfreq, x^2, x^3, x^7, x^8 , x]

Fig. 6.8 *Mathematica* histogram of the frequencies of the building heights

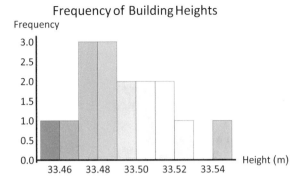

Frequency of Building Heights

Out[27]:= $0.559748x^2 - 0.099054x^3 + 0.0000232269x^7 - 1.93489 * 10^{-6}x^8$

Plot of the 'fit' curve

```
In[28]:= BCURVE = Plot[bestcurve, {x, 0, 9.65},
PlotStyle -> {Blue, Thick}, AxesLabel ->
{HoldForm["Height (m)"], HoldForm
["Frequency"]},
PlotLabel -> HoldForm["Frequency of Building
Heights" ], LabelStyle -> {FontFamily ->
"Calibri", 12, GrayLevel[0]},
PlotRange -> {{0, 9.65}, {0, 3.2}}, AxesStyle
-> Thick,
TicksStyle -> Directive[Thin]];
```

The *Show* command can be used to superimpose the frequency 'fit curve' on to the bar chart

```
In[29]:= QQ6=Show[P1, BCURVE]
]
```

The combined bar chart and 'fit' curve (*QQ6*) is shown in Fig. 6.13.

6.1.4 Solution of Example 1 Using Matlab

Matlab takes a bit more learning than *Maple* or *Mathematica*, but the extra effort is well worth it. Each of the three software packages has its advantages. For *Matlab*, the advantage is that it allows a variety of mathematical operations to be carried out on arrays, and is hence ideal for matrix operations. We will assume some familiarity with *Matlab* features and offer only a brief overview.

When *Matlab* is opened, four windows appear together with a top menu bar that leads to a number sub-menu bars offering comprehensive choices. The four win-

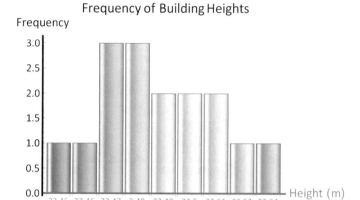

Fig. 6.9 *Mathematica* bar chart of the frequencies of the building heights

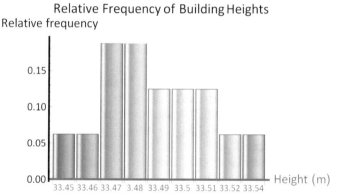

Fig. 6.10 *Mathematica* bar chart of the relative frequencies of the building heights

Fig. 6.11 *Mathematica* line and data plot of the frequencies of the building heights

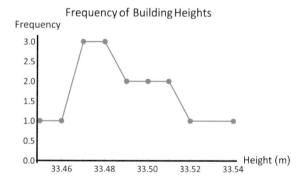

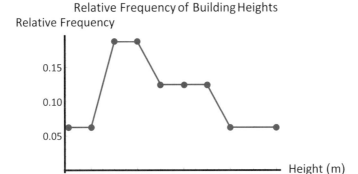

Fig. 6.12 *Mathematica* line and data plot of relative frequencies of the building heights

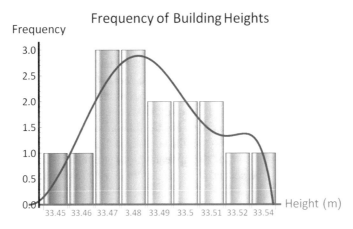

Fig. 6.13 *Mathemtica* bar chart of the frequencies of the building heights together with a fitted frequency curve

dows are: *Current Folder*, *Workspace*, *Editor* and *Command Window*. A window is invoked by left-clicking the mouse button on the window of interest. Commands placed in the *Command Window* are executed immediately by pressing the keyboard *return* key. This is fine for a few simple commands. But, when a large number of commands are involved, and especially to keep a permanent record of the command sequences, writing the commands in the editor window then saving them to a *Matlab* .m file is much preferred. The sequence of commands is, of course, a program. To run the program, choose the *Editor* sub-menu bar from the top menu bar and left-click the *Run* icon. The output appears in the command window. Plots are displayed in a separate graphics window which opens after the program is ran.

Help on any *Matlab* command or option is obtained by using the *help* command in the *Command Window*. For example *help errorbar* gives help on the *errorbar* command used, as the name implies, to produce error bars on a graph.

In *Matlab* there are two sets of operators for multiplication, division and exponentiation. The ordinary symbols $*$, $/$ and $^\wedge$ are reserved for matrix operations. The 'dot' versions $.*$, $./$ and $.^\wedge$ are reserved for element-by-element operations between arrays and for ordinary multiplication of functions. Since numbers can be viewed as scalar matrices both forms of the operations can be used with numbers. For addition and subtraction, no such distinction is necessary and so the usual $+$ and $-$ symbols are used for all cases. Some examples are:

$$[1, 2, 3]. * 3 = [3, 6, 9], \quad [12, 9, 15]./3 = [4, 3, 5], \quad [1, 2, 3].^2 = [1, 4, 9],$$

$$3. * 4 = 12, \quad \text{or} \quad 3 * 4 = 12, \quad \text{both forms are equally valid}$$

$\sin(x). * \cos(x), \quad$ The correct way to multiply functions

$\sin(x) * \cos(x), \quad$ The wrong way to multiply functions. Results in an error message

There are two types of *Matlab* .m files: the *script* .m file and the *function* .m file. As its name implies, the latter is used to define functions. To write a program, the *script* option is selected in the editor window by left-clicking the *New* icon in the *Editor* sub-menu, then left-clicking the *script* option. Once the program is written in the *Editor* window it is saved as a *script* .m file when the *Save* icon or option is selected. The *Function* definition is selected in the same way except that the *function* option is selected instead. In this case, a function definition template appears in the *Editor* window. This is a great help since functions must be defined with a specific format and it is this format that distinguishes the *function.m* file from a *script.m* file. We will come across *function.m* files in the solution program.

There are two ways to control the appearance of graphics. The first is to use command line options. The second is to use *Matlab*'s graphical interface. We will use the command line options to produce the charts and plots that follow as this is more instructive. But, the graphical interface is very useful and so we will describe how to invoke it.

When a program containing a chart and/or plot is ran, a new window opens to display the chart or plot. The new window also contains a menu bar which allows a variety of options that control the appearance of the chart or plot to be chosen. The steps to invoke the graphical interface are as follows:

(1) In the new graphics window displaying the plot or chart, left-click on the *Tools* option in the top menu bar to open the *Tools* menu .
(2) Left-click on the *Edit plot* option.
(3) Double left-clicking on the background of the plot brings up the *Property Editor* window specific to the background. Here, a title and axis labels can be entered, a number of grid configurations can be chosen, a background colour can be selected and so on. On the other hand, double clicking on the bars of a chart, the points of a data plot, or the curve of a line plot brings up the *Property Editor* window specific to these graphic elements.

Curve fitting can be done either with *Matlab*'s *polyfit* command or with a graphical interface. The *polyfit* command produces a best fit polynomial function which can then be plotted. An explanation for doing this is given in the solution program.

Invoking the graphical curve fitting interface is achieved by following the same initial steps as for graphics, except that *Basic Fitting* is selected from the *Tools* menu. A menu window appears offering a choice for the degree of the polynomial to be used for the curve fitting. Once the choice is made the curve is plotted on the data points graph or on the chart. The numerical values of the coefficients of the polynomial used for the curve fitting are returned in same options window.

The *Matlab* program for the Solution of Example 1 follows:

```
% MATLAB PROGRAM FOR THE SOLUTION OF EXAMPLE 1. MEASURED
% BUILDING HEIGHTS

clear
format long
% 'Format long' specifies a scaled fixed point format with 15 digits for double
% precision (and 7 digits when single precision is specified with, for example, 'Y
% = single(X)' ). Other format options can be found by typing 'help format' in the
% 'Command Window'

% xxxxxxxxxxxxxxxxxxxxxxxxxxxxxxxxxxxxxxxxxxxxxxxxxxxxxxxxxxxxxxxxxxxxxxxx
% Matlab is framed in terms of arrays, mostly numerical arrays (matrices). In view
% of this, what we have been calling lists in Maple or Mathematica are better referred
% to as one-dimensional matrices, i.e., vectors. Therefore, we refer to our sets of
% measured values as vectors rather than lists.
% xxxxxxxxxxxxxxxxxxxxxxxxxxxxxxxxxxxxxxxxxxxxxxxxxxxxxxxxxxxxxxxxxxxxxxxx

% The percent sign % indicates a text line. Commands can be temporarily suppressed
% by converting them to text using the % sign.

% The data points are entered as a vector labelled 'bheights':
bheightsi = [33.45, 33.46, 33.47, 33.50, 33.49, 33.51, 33.48, 33.52, 33.47, 33.48, ...
33.49, 33.47, 33.51, 33.50, 33.48, 33.54]
% The ellpsis '...' mean that the command is to be continued on the next line

% Arrays of the frequencies and relative frequencies:
bheightsfreq=[1,1,3,3,2,2,2,1,1]
bheightsrelfreq=[1,1,3,3,2,2,2,1,1]/16

% xxxxxxxxxxxxxxxxxxxxxxxxxxxxxxxxxxxxxxxxxxxxxxxxxxxxxxxxxxxxxxxxxxxxxxxx
% One of the advantages of Matlab is that it allows many mathematical operations
% to be applied to the whole array. In the above command each element of the array
% is divided by 16.
% xxxxxxxxxxxxxxxxxxxxxxxxxxxxxxxxxxxxxxxxxxxxxxxxxxxxxxxxxxxxxxxxxxxxxxxx

% The 'sort' command is used to place the measurements in ascending order
bheights=sort(bheightsi)
```

% The 'sum' command is used to sum the measurements
bheightssum=sum(bheights)

% Dividing by the total number 16 of measurements gives the mean
bheightsmean=bheightssum/16

% Subtracting each measurement from the mean gives the residuals
bheightsresiduals=bheights-bheightsmean

% The following is a check that 'bheightresiduals' sums to zero
bheightsrescheck=sum(bheightsresiduals)

% xxxxxxxxxxxxxxxxxxxxxxxxxxxx NOTE xxxxxxxxxxxxxxxxxxxxxxxxxxxxxxxxx
% 'bheightsrescheck' should be zero, but numerical error gives a very small number
% xxxxxxxxxxxxxxxxxxxxxxxxxxxxxxxxxxxxxxxxxxxxxxxxxxxxxxxxxxxxxxxxxx

% Calculation of the squares of the residuals:
 bheightsresidualssq=bheightsresiduals.^2

% Determination of the sum of the residuals squared
bheightsressqsum=sum(bheightsresidualssq)

% Taking the mean of 'bheightsresiduesum' gives the variance
bheightsvar=bheightsressqsum/16

% The square root of the variance gives the standard deviation
bheightsstdv=sqrt(bheightsvar)

% Determination of the standard error in the mean:
bheightstanerror=bheightsstdv/sqrt(16-1)

% xxxxxxxxxxxxxxxxxx ANSWER xxxxxxxxxxxxxxxxxxxxxxxxxxxxxxxxxxxxxxxxx
%
% Height of the building = 33.49 p/m 0.01 m
%
% The original measurements were made to 4 significant figures (sf) so that the
% calculated mean cannot be more accurate than this. Hence, the standard error in
% the mean must be rounded to 2 decimal places so that it corresponds to the 4th sf
% of the mean height. The answer is therefore given to 4 sf with the error of the 4th
% sf indicated.
% xxxxxxxxxxxxxxxxxxxxxxxxxxxxxxxxxxxxxxxxxxxxxxxxxxxxxxxxxxxxxxxxxx

% xxxxxxxxxxxxxxxxxxxxxxxxx INBUILT FUNCTIONS xxxxxxxxxxxxxxxxxxxxxxx
% Matlab has inbuilt functions for the mean, variance and standard deviation, namely,
% 'mean ()', 'var ()' and 'std ()', respectively. But, it should be noted that the latter

% two functions deliver the sample variance and sample standard deviation. We have
% preferred to enter the formulae by hand, first, because it is more instructive, and
% second, because it is the standard deviation that is needed to calculate the standard
% error in the mean, not the sample standard deviation.
% xxxxxxxxxxxxxxxxxxxxxxxxxxxxxxxxxxxxxxxxxxxxxxxxxxxxxxxxxxxxxxxxx

% xxxxxxxxxxxxxxxxxxxxxxxxxxxxxxxxxxxxxxxxxxxxxxxxxxxxxxxxxxxxxxxxx
% xxxxxxxxxxxxxxxxxxxxxxxxxxxxx THE CHARTS xxxxxxxxxxxxxxxxxxxxxxxxx
% xxxxxxxxxxxxxxxxxxxxxxxxxxxxxxxxxxxxxxxxxxxxxxxxxxxxxxxxxxxxxxxxx

% There are a variety of options for controlling the appearance of charts and graphs.
% In Matlab, the meaning and use of graphical options is usually obvious from their
% use in the command line. Similarly, the syntax of the options is fairly clear from
% their use in the command line. Hence, we will only add a few extra explanatory
% comments here and there. More options and values of the options can be found
% by using Matlab's 'help' command as described in the introduction of the main
% text of this section.

% xxxxxxxxxxxxxxxxxxxxxxxxxxxxx BAR CHARTS xxxxxxxxxxxxxxxxxxxxxxxxx
% With some exceptions, the graphics options apply to all types of charts and graphs.
% The options we present are, of course, by no means comprehensive. Note that
% Matlab has separate commands for histograms and bar charts. Although height is
% a continuous quantity, the measured values are necessarily discrete. Bar charts
% are preferred in the case of discrete values because the column labels refer to the
% discrete value and the frequency also refers to that value. Histograms tend to count
% data points in unequal intervals, Another advantage of bar charts is that the plots
% of relative frequencies can be produced in a straight forward way. We will,
% however, include histograms by way of example.

% xxxxxxxxxxxxxxxxxxxxxxxxxxxxx NOTE xxxxxxxxxxxxxxxxxxxxxxxxxxxxxxx
% The x-axis definition is only needed for the 'plot' command. The 'bar' command
% numbers the x-axis automatically.

% xxxxxxxxxxxxxxxxxxxxxxxxxxxxx ARRAYS OF PLOTS AND CHARTS xxxxxxxxxxxx
% The 'subplot(m,n,p)' command produces an array of plots with m rows and n
% columns. The position of the plot in the array is specified by p. The 'plot' command
% produces a single graphic. A number of options are given by enclosing an
% indicator with apostrophes. The options may be line types, point types, or colours.
% For example, the indicator for cyan is c so that 'c' in a 'bar' command colours the
% bars cyan, while 'b' in a 'plot' command colours the line blue. The following is a
% list of indicators, not all of which can be used in the 'bar' or 'hist' commands, but
% can be used in the 'plot' command:

% xxxxxxxxxxxxxxxxxxxxxxxxxxxxx LINE INDICATORS xxxxxxxxxxxxxxxxxxxxxxxx
% solid = '-' (is the default value so that a solid line is plotted in the absence of an

```
% option), dotted = ':', dash-dot = '-.', dashed = '--'
% xxxxxxxxxxxxxxxxxxxxxxxxx DATA POINT INDICATORS xxxxxxxxxxxxxx
% point = '.', circle = 'o', x-mark = 'x', plus = '+', star = '*', square = 's',
% diamond ='d', triangle down = 'v', triangle up = '^', triangle left = '<',
% triangle right = '>', pentagram = 'p', hexagram = 'h'


% xxxxxxxxxxxxxxxxxxxxxxxxxxxxxxxx COLOUR xxxxxxxxxxxxxxxxxxxxxxxxxx
% blue = 'b', green = 'g', red = 'r', cyan = 'c', magenta = 'm', yellow = 'y',
% black = 'k', white = 'w'
% xxxxxxxxxxxxxxxxxxxxxxxxxxxxxxxxxxxxxxxxxxxxxxxxxxxxxxxxxxxxxxxx


% xxxxxxxxxxxxxxxxxx DEFINITION OF THE x, y1 and y2 AXES xxxxxxxxxxxxx
% 'bhieghtsfreq' and 'bheightsrelfreq' can be used directly in the 'plot' command,
% but relabelling them by y1 and y2,respectively, makes the plot command look
% neater.
x=[1:1:9]
% The x vector labels the position of the height value in the list 'bheights', with equal
% values counted in the same position. There are two reasons for labelling the x-axis
% in this way: First, labelling the x-axis with the four significant figure heights
% clutters the x-axis, with numbers overlapping in some cases. Second, the 'bar'
% command numbers the x-axis automatically and so corresponds to the positions
% of the heights in the list 'bheights' with equal values counted in the same position.
y1=bheightsfreq
y2=bheightsrelfreq
% The syntax [1:1:9] produces a one-dimensional array of numbers 1 to 9 in steps
% of 1 against which y1 and y2 is plotted.


% xxxxxxxxxxxxxxxxxxxxxxxxxxxxxxx BAR CHARTS xxxxxxxxxxxxxxxxxxxxxxxxxxx
% Bar chart of the frequencies of the measured building heights labelled by their
% position in 'bheights' (with equal values counted in the same position)
subplot(2,2,1)
bar(y1, 'c'),title('Frequency of Building Heights'),
xlabel('Height position'), ylabel('Frequency'), axis([0,10,0,3.5])


% Bar chart of the relative frequencies of the measured building heights labelled by
% their position in 'bheights' (with equal values counted in the same position)
subplot(2,2,2)
bar(y2,'m'),title('Relative Frequency of Building Heights'),
xlabel('Height position'), ylabel('Relative frequency'),
axis([0,10,0,0.25])


% zzzzzzzzzzzzzzzzzzzzzzzzzzzzzzz LINE PLOT xxxxxxxxxxxxxxxxxxxxxxxxxxxxx
% The following syntax in the 'plot' command plots two or more curves or data
% points on the same axes, and controls the appearance of each:
% plot (x1, y1, indicator1, x2, y2, indicator2, ...)
```

% We will use this form of 'plot' to combine curves and data points on the same axes.

% Plot of joined frequency data points of the measured building heights labelled
% by their position in 'bheights' (with equal values counted in the same position)
subplot(2,2,3)
plot(x,y1,'p',x,y1),title('Frequency of Building Heights'),
xlabel('Height position'), ylabel('Frequency'), grid, axis([0,10,0,3.5])
% The indicator 'p' in the first occurrence of x, y1 results in the data points being
% represented by a pentogram symbol. The second occurrence of the same data
% points x,y1 results in a straight line. The indicator '-' for a solid line is not needed
% since, as mentioned above, it is the default value.

% Plot of joined relative frequency data points of the measured building heights
% labelled by their position in 'bheights' (with equal values counted in the same
% position)
subplot(2,2,4)
plot(x,y2,'o',x,y2), title('Relative Frequency of Building Heights'),
xlabel('Height position'), ylabel('Relative frequency'), grid,
axis([0,10,0,0.25])

% The four graphs are shown in Fig. 6.14.

% xxxxxxxxxxxxxxxxxxxxxxxxxx HISTOGRAM xxxxxxxxxxxxxxxxxxxxxxxxxxxxxxxx
% As mentioned in the main text of this section, bar charts are preferred to display
% the frequency and relative frequency of the measured building heights. But, by
% way of example, we give the syntax for plotting a histogram of the height
% frequencies, but we will not display the result (by suppressing the command using
% the percent sign %).

% hist(bheights,9),title('Frequency of Building Heights')
% xlabel('Height position'), ylabel('Frequency')

% xxxxxxxxxxxxxxxxxxxxxxxxxxxxx CURVE FITTING xxxxxxxxxxxxxxxxxxxxxxxxxx
% Using the 'polyfit' command, a curve can be fitted to the frequencies of the heights
% in the vector 'bheights'

 polyfit(x,y1,4)

% The syntax is:
% polyfit(x,y,n)- x is vector of x values, y is a vector of y-values and n is the degree
% of the polynomial used for the curve fitting. E.g., n=4 is the fourth degree
% polynomial $ax^4+bx^3+cx^2+dx+e$. *polyfit* returns numerical values of
% *a*, *b*, *c*, *d* and *e* which produce the best fit curve.

% A polynomial fit function can be defined using the results from 'polyfit', then

% saved as a Matlab function file named FREQFIT.m

% The plot of the function FREQFIT against the positions of the building heights
% in 'bheights' (with equal values assigned separate positions)
plot(x,FREQFIT(x),x,y1,'p'), axis([0,10,0,3.5]),
title('Building Heights'), xlabel('Position in bheights'),...
ylabel('Frequency')

% The fitted curve and the data points are shown in Fig. 6.15.

% xxxxxxxxxxxxxxxxxxxxxxxxxxx ERROR BARS xxxxxxxxxxxxxxxxxxxxxxxxxxxxxxx
% Finally, we want to show how error bars can be included using the 'error bar'
% command with syntax:
% errorbar(x,y,dy) - plots y-values against x-values with error bars (y-dy,y+dy).

% The plot with error bars is shown in Fig. 6.16.

% There are variations in the syntax of the 'errorbar' command which can be seen
% by typing 'help errorbar' in the command line of the command window.

% Definition of the z vector and error bar vector e: The two vectors must have the
% same dimensions. The error vector contains the error in each reading as given by
% the standard error in the mean. In our case. the standard error is 0.01 and is the
% same for each measurement.
% First define the list z of building heights (i.e. relabel 'bheights' by z to simplify
% the look of the 'errorbar' command), and define a list e of errors of the same length
% as 'bheights
z=bheights
e=[0.01, 0.01, 0.01, 0.01, 0.01, 0.01, 0.01, 0.01, 0.01, 0.01, 0.01, ...
0.01, 0.01, 0.01, 0.01, 0.01]

% Plot of the building heights versus their position in 'bheights' (with equal values
% assigned separate positions) together with error bars, i.e., plot z versus x1 together
% with error bars
errorbar(z, e), title('Building Heights'),...
xlabel('Position in bheights'), ylabel('Height (m)')

% Note that Matlab overwrites earlier graphics when more than one graphic or
% graphic array is included in a program. The easiest way to see the plot or plot array
% of current interest is to temporarily suppress other plots or plot arrays using the
% percent sign %.

% xxxxxxxxxxxxxxxxxxxxxxxxxx END OF MATLAB PROGRAM xxxxxxxxxxxxxxxxx

The fit function used in the *plot* command to plot the fitted curve with the data points

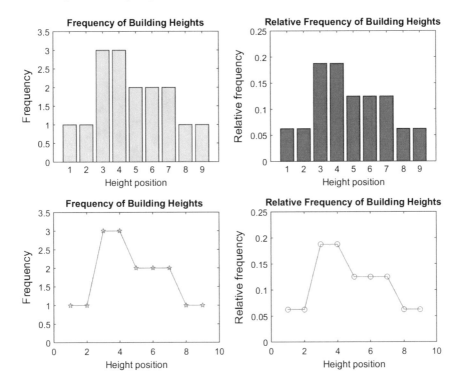

Fig. 6.14 *Matlab* plots. TOP: Bar charts of the frequencies and relative frequencies of the building heights labelled by their position in *bheights*. BOTTOM: Joined data points of the frequencies and relative frequencies of the building heights also labelled by their position in *bheights*

is defined in the following *Matlab*.m function file named FREQFIT.m. Note that the syntax of the top and bottom lines define the file to be a function file as opposed to a program file:

```
[−] function [FRFT] = FREQFIT(x)
 | % Matlab polynomial fit to the frequencies of the heights in 'bheights'
 | FRFT=0.003496503496504.*x.^4-0.051411551411553.*x.^3+ ...
 |0.086441336441346.*x.^2+ 0.974682724682710.*x-0.222222222222230
 └end
```

6.1.5   *Comments on the Solution Programs of Example 1*

The answer from all four programs is:

Answer. Height of the building = 33.49 ± 0.01 m

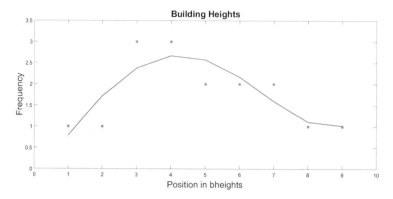

Fig. 6.15 The graph shows the fit curve produced by the polynomial $ax^4 + bx^3 + cx^2 + dx + e$ using the values of the coefficients produced by *Matlab*'s *polyfit* command

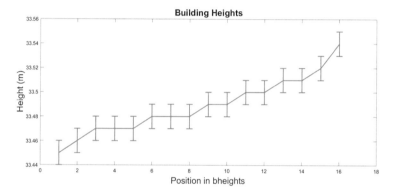

Fig. 6.16 *Matlab* plot of the measured building heights labelled by their position in 'bheights' together with error bars

The result shows reasonable precision. Since we judge the systematic error to be negligible compared to the random error, we may also conclude that the measurement is reasonably accurate.

The charts and joined data plots of the frequencies approach a Gaussian distribution as is expected for this kind of measurement, though the last quarter of the curve shows a marked deviation from a Gaussian curve. This is probably because the measurements were too few.

The spreadsheet format of *Excel* is convenient for calculations of the type carried out above. In *Excel*, the plotting of bar charts and data line graphs is automated, producing quality results very simply. Various display options are offered. The mathematical packages require more work, but offer a large range of display options for charts and graphs. For plotting functions, such as the best straight line, the mathematical packages are much preferred.

6.2   Example 2. Acceleration due to Gravity

Example 2 is an experiment to determine the acceleration due to gravity by measuring the time t for an object to fall through various heights. An object is dropped from various heights, measured with a ruler, and the time of fall is measured with a stop watch. Five times are taken for each height. The results are as follows:

Height h (m)	Time of fall t (seconds)				
1.1	0.475	0.470	0.470	0.471	0.477
1.0	0.456	0.451	0.455	0.449	0.454
0.9	0.431	0.433	0.435	0.427	0.421
0.8	0.395	0.412	0.406	0.397	0.405
0.7	0.376	0.364	0.375	0.382	0.381
0.6	0.351	0.347	0.348	0.352	0.346
0.5	0.320	0.323	0.316	0.318	0.326
0.4	0.282	0.276	0.285	0.277	0.283
0.3	0.250	0.255	0.246	0.251	0.247
0.2	0.202	0.197	0.194	0.209	0.206
0.1	0.153	0.133	0.149	0.151	0.131

The acceleration due to gravity g is found using equation,

$$h = \frac{1}{2}gt^2. \tag{6.1}$$

Here, we want to find g graphically. Casting the equation in the form

$$\frac{2}{g} = \frac{t^2}{h} \tag{6.2}$$

shows that g can be determined from the slope of a linear graph (invariably preferred to nonlinear graphs) of t^2 versus h, i.e.,

$$\frac{2}{g} = \text{slope}$$

$$g = \frac{2}{\text{slope}} \tag{6.3}$$

Since, as seen from Eq. (6.1), for $t = 0$, $h = 0$ the line passes through the origin. The method of least squares offers two approaches for finding the best straight line for this case:

Approach 1: Ignore this information and determine both the best slope m and the best c and their errors using Eqns. (4.5), (4.6) and (4.10)–(4.13). This approach has the advantage of visually indicating the size of an error by the distance the line misses the origin. For small errors, the distance of the line from the origin may be too small to be visible.

Approach 2: Use the information and use formulae (4.14) to (4.16) for lines through the origin to determine the best slope m and its error, then draw the line with this slope through the origin.

We want to do the following:

1. Use approach 1 to determine the slope m of the best line and where the best line cuts the y-axis, together with their errors.
2. Use the best value for the slope m and its error Δm to calculate the acceleration due to gravity g and its error.
3. Plot the equation of the line with the best slope m passing through c and add the data points defined by the heights and the mean times squared for each height.
4. Use approach 2 to determine the slope m of the best line through the origin together with its error.
5. Define the equation of the line with best slope m and $c = 0$, and hence plot the best straight line through the origin.
6. Repeat the calculation of 2 for the m of the line through the origin.
7. Combine the plots of the best straight lines from the two approaches and the data points in various combinations.

The aim of Example 2 is to show how to use the four packages to perform calculations using the method of least squares and then to plot the best straight lines and data points.

6.2.1   Excel Solution of Example 2. Determination of the Acceleration due to Gravity

Here we assume that the *Excel* spreadsheet solution of example 1 has been studied, so we won't repeat the explanations given there. The *Excel* spreadsheet solution of Example 2 follows the same pattern as for Example 1, only the formulae entered in the formula cells are different. We will therefore give the hidden formula in the formulae cells.

Since times were taken to 3 sf, calculations in the spreadsheet were carried out to 4 sf.

The mean values of the 5 times squared, $\langle t \rangle_i^2$, for each height will be used to plot the graph. Therefore, in formulae (4.5), (4.6), (4.10)–(4.13) and (4.14)–(4.16) the heights h_i correspond to x_i, and the $\langle t \rangle_i^2$ corresponds to y_i.

FIRST. Determination of the best m and c and their errors using formulae (4.5), (4.6) and (4.10) to (4.13)

	A	B	C	D	E	F	G	H	I	J	K	L	M	N
1		FIRST. Determination of best m and c and their errors using formulae (4.5), (4.6) and (4.10) to (4.13)												
2		Height $x_i = $ h_i (m)	Time t of fall (s)					Sum of t	$<t>\_i$	$y\_i =$ $<t>\_i^2$	$(x\_i-<x>) =$ $(h\_i-<h>)$	$(x\_i-<x>)^2$ $= (h\_i-<h>)^2$	$(x\_i-<x>)y\_i =$ $(h\_i-<h>)<t>\_i^2$	$d\_i^2 =$ $(y\_i-mx\_i-c)^2$
3	1	1.1	0.475	0.470	0.470	0.471	0.477	2.3630	0.4726	0.2234	0.5000	0.2500	0.111675	0.00000059481
4	2	1.0	0.456	0.451	0.455	0.449	0.454	2.2650	0.4530	0.2052	0.4000	0.1600	0.082084	0.00000221111
5	3	0.9	0.431	0.433	0.435	0.427	0.421	2.1470	0.4294	0.1844	0.3000	0.0900	0.055315	0.00000112857
6	4	0.8	0.395	0.412	0.406	0.397	0.405	2.0150	0.4030	0.1624	0.2000	0.0400	0.032482	0.00000026319
7	5	0.7	0.376	0.364	0.375	0.382	0.381	1.8780	0.3756	0.1411	0.1000	0.0100	0.014108	0.00000209282
8	6	0.6	0.351	0.347	0.348	0.352	0.346	1.7440	0.3488	0.1217	0.0000	0.0000	0.000000	0.00000021213
9	7	0.5	0.320	0.323	0.316	0.318	0.326	1.6030	0.3206	0.1028	-0.1000	0.0100	-0.010278	0.00000112857
10	8	0.4	0.282	0.28	0.285	0.277	0.283	1.4030	0.2806	0.0787	-0.2000	0.0400	-0.015747	0.00000668563
11	9	0.3	0.250	0.255	0.246	0.251	0.247	1.2490	0.2498	0.0624	-0.3000	0.0900	-0.018720	0.00000218455
12	10	0.2	0.202	0.197	0.194	0.209	0.206	1.0080	0.2016	0.0406	-0.4000	0.1600	-0.016257	0.00000001453
13	11	0.1	0.153	0.133	0.149	0.151	0.131	0.7170	0.1434	0.0206	-0.5000	0.2500	-0.010282	0.00000019496
14	$<h> \geq$ 0.6000								$<<t>\_i^2> \geq$ 0.12211		$D =$	1.1000		
15									Sum of $(x\_i-<x>) y\_i = (h\_i - <h>)<t>\_i^2 =$			0.224379		
16												Sum of $d\_i^2 =$	0.000016711	

Best slope m =	0.2040
Standard error in slope m =	0.001299
Standard error in g =	0.06245
Standard error in g to 1 sf =	0.06
$c = <y> - m<x> = <<t\_i>^2> - m<h> =$	-0.0002780
Standard error in c =	0.0008812
$g = 2/m =$	9.8048

Answer. Acceleration due to gravity is $g = 9.80 \pm 0.06$ m/s^2

Fig. 6.17 *Excel* spreadsheet to find g by first determining the best m and c and their errors using formulae (4.5), (4.6) and (4.10) to (4.13)

For what follows, refer to Fig. 6.17. Cells A3 to A13 count the data. The heights h_i are entered in column B. The five times measured for each height are placed in columns C to G. Aside from text cells the remaining cells contain formulae. The formulae are hidden; only the results of the formula are shown in the formulae cells. We will therefore list the formulae cells together with the hidden formula they contain by first giving the cell address followed by the hidden formula it contains (enclosed in square brackets). We will also add a description of the formula. For columns containing cells with the same formula referring to different rows we will give the formula in the first cell of the column and indicate that the formula is repeated in the remaining cells with the abbreviation FRF (formula repeated from) followed by 'cell i → cell f'.

- B14: [=SUM(B3:B13)/11], $\langle h \rangle$
- H3: [=SUM(C3:G3)], sum of the five times for h_1, FRF H4 → H13
- I3: [=H3/5], $\langle t \rangle_1$, FRF I4 → I13
- J3: [=I3*I3], $\langle t \rangle_1^2$, FRF J4 → J13
- J14: [=SUM(J3:J13)/11], $\langle \langle t \rangle_i^2 \rangle$
- K3: [=B3-0.6], residual of h_i, FRF K4 → K13
- L3: [=K3*K3], square of the residual h_1, FRF L4 → L13
- L14: [=SUM(L3:L13)], D

- M3: [=K3*J3], $(h_1 - \langle h \rangle)\langle t \rangle_1^2$, FRF M4 → M13
- M15: [=SUM(M3:M13)], $\sum_{i=1}^{11}(h_i - \langle h \rangle)\langle t \rangle_i^2$
- N3: [=(J3-0.204*B3-(-0.000278))^2]
- N16: [=SUM(N3:N13)], $\sum_{i=1}^{11} d_i^2$
- I17: [= M15/L14], best m
- I18: [=SQRT(N16/L14/(11-2))], Δm
- I19: [= I23*I18/I17], Δg
- I20: [= I19], Δg rounded to 1 sf.
- I21: [=J14-I17*B14], c
- I22: [=SQRT((1/11+B14*B14/L14)*N16/(11-2))], Δc
- I23: [=2/I17], g

The answer from the *Excel* spreadsheet in Fig. 6.17 obtained by approach 1 is:

Acceleration due to gravity is $g = 9.80 \pm 0.06$ ms^{-2}

Since the 1st (rounded) significant figure of the standard error in g corresponds to the 3rd significant figure of g, the calculated value of g is given to 3 sf.

Comparing our result with the value of $g = 9.81$ found in tables of physical constants, established by much more sophisticated experiments, we see that our simple experiment has produced a reasonable value of g with the accurate value from tables lying within our error limits. We will comment on the result further in Sect. 6.2.5.

SECOND. Determination of the best m and its error using formulae (4.14)–(4.16) for a line through the origin

For what follows refer to Fig. 6.18. The cells of column A count the data. The heights h_i are entered in column B, while column C contains the values of $\langle t \rangle_i^2$. The remaining cells contain either formulae or text. As with the spreadsheet of Fig. 6.17, formulae are hidden, so we will list the contents of formulae cells using the same format as for the spreadsheet of Fig. 6.17:

- B14: [=SUM(B3:B13)/11], $\langle h \rangle$
- C15: [=SUM(C3:C13)/11], $\langle \langle t \rangle_i^2 \rangle$
- D3: [=B3*B3], h_i^2, FRF D4 → D13
- D16: [=SUM(D3:D13)], $\sum_{i=1}^{11} h_i^2$
- E3: [=B3*C3], $h_i \langle t \rangle_i^2$, FRF E4 → E13
- E17: [=SUM(E3:E13)], $\sum_{i=1}^{11} h_i \langle t \rangle_i^2$
- F3: [=(C3-B3*0.2036)^2], d_i^2, FRF F4 → F13
- F18: [=SUM(F3:F13)], $\sum_{i=1}^{11} d_i^2$
- F20: [E17/D16], best slope m
- F21: [=SQRT(F18/D16)/(11-1)], Δm
- F22: [=F24*F21/F20], Δg
- F23: [=F22], Δg rounded to 1 sf.
- F24:[=2/F20], g

	A	B	C	D	E	F
1			**SECOND. Determination of best m and its error using formulae (4.14) to (4.16) for a line through the origin**				
2		Height $x\_i = h\_i$ (m)	$y\_i = <t>\_i{}^{\wedge}2$	$x\_i{}^{\wedge}2 = h\_i{}^{\wedge}2$	$x\_i * y\_i =$ $h\_i * <t>\_i{}^{\wedge}2$	$d\_i{}^{\wedge}2 = (y\_i - mx\_i) =$ $(<t>\_i{}^{\wedge}2 - m\,h\_i){}^{\wedge}2$
3	1	1.100	0.2234	1.2100	0.2457	0.00000031360
4	2	1.000	0.2052	1.0000	0.2052	0.00000256000
5	3	0.900	0.1844	0.8100	0.1660	0.00000134560
6	4	0.800	0.1624	0.6400	0.1299	0.00000023040
7	5	0.700	0.1411	0.4900	0.0988	0.00000201640
8	6	0.600	0.1217	0.3600	0.0730	0.00000021160
9	7	0.500	0.1028	0.2500	0.0514	0.00000100000
10	8	0.400	0.0787	0.1600	0.0315	0.00000750760
11	9	0.300	0.0624	0.0900	0.0187	0.00000174240
12	10	0.200	0.0406	0.0400	0.0081	0.00000001440
13	11	0.100	0.0206	0.0100	0.0021	0.00000005760
14	$<h\_i$	0.600				
15	$<<t>\_i{}^{\wedge}2>=$		0.1221			
16			Sum of $h\_i{}^{\wedge}2=$	5.0600		
17				Sum of $h\_i * <t>\_i{}^{\wedge}2=$	1.0304	
18					Sum of $d\_i{}^{\wedge}2=$	0.0000170
19						
20					Best slope m =	0.20363439
21					Standard error in slope m =	0.00057962
22					Standard error in g =	0.02795580
23					Standard error in g to 1 sf =	0.03
24					$g = 2/m$ =	9.8215
25						
26		**Answer. Acceleration due to gravity is g = 9.82 ± 0.03 m/s^2**				

Fig. 6.18 *Excel* spreadsheet to find g by first determining the best m and its error using formulae (4.14) to (4.16) for a line through the origin

The answer from the *Excel* spreadsheet in Fig. 6.17, obtained by approach 2, is:

Acceleration due to gravity is $g = 9.82 \pm 0.03$ ms^{-2}

Since the 1st (rounded) significant figure of the standard error in g corresponds to the 3rd significant figure of g, the calculated value of g is given to 3 sf.

We will comment on the apparent extra accuracy using approach 2 in the overall comment section, Sect. 6.2.5, for Example 2.

6.2.2   Maple Program for the Solution of Example 2.
Determination of the Acceleration due to Gravity

The *Maple* program for the solution of Example 2 follows. For introductory aspects on the *Maple* language, see subsection 6.1.2 in which *Maple* is used for Example 1. Additional commands that are needed for Example 2 are explained in the solution program that follows.

[*MAPLE* **PROGRAM FOR THE SOLUTION OF**

EXAMPLE 2. To determine the acceleration due to

gravity by measuring the time *t* **for an object to fall**

through various heights h **and by using the methodof least squares**

[**APPROACH 1. Determination of the best** m **and**
c and their errors using formulae (4.5), (4.6) and
(4.10) to (4.13)

[

[A colon placed at the end of a command line suppresses output. A semi-colon at the end of the command line or no punctuation at the end of the command line shows output. You can choose a cell to be a text cell by pressing the 'T' button in the menu bar. A new input cell is chosen either by pressing return after an existing input cell, or by pressing the button '[>'.

[

[To access plotting, curve fitting and histogram plotting, the packages *plots,   CurveFitting* and   *Statistics*   must   be   loaded   with   the *with*(*package name*) command

[> *restart*

    [The *restart* command clears memory. It is always a good idea to start a program with this command

[> *with*(*plots*):

[> *with*(*Statistics*):

[> *with*(*CurveFitting*):

[

[We will use both arrays and lists, whichever is more convenient for a particular calculation.

[

	A	B	C	D	E	F
1		**SECOND. Determination of best m and its error using formulae (4.14) to (4.16) for a line through the origin**					
2		Height $x\_i = h\_i$ (m)	$y\_i =$ $<t>\_i{\char`\^}2$	$x\_i{\char`\^}2 = h\_i{\char`\^}2$	$x\_i{*}y\_i =$ $h\_i{*}<t>\_i{\char`\^}2$	$d\_i{\char`\^}2 =(y\_i{-}mx\_i)$ = $(<t>\_i{\char`\^}2{-}m\,h\_i){\char`\^}2$
3	1	1.100	0.2234	1.2100	0.2457	0.00000031360
4	2	1.000	0.2052	1.0000	0.2052	0.00000256000
5	3	0.900	0.1844	0.8100	0.1660	0.00000134560
6	4	0.800	0.1624	0.6400	0.1299	0.00000023040
7	5	0.700	0.1411	0.4900	0.0988	0.00000201640
8	6	0.600	0.1217	0.3600	0.0730	0.00000021160
9	7	0.500	0.1028	0.2500	0.0514	0.00000100000
10	8	0.400	0.0787	0.1600	0.0315	0.00000750760
11	9	0.300	0.0624	0.0900	0.0187	0.00000174240
12	10	0.200	0.0406	0.0400	0.0081	0.00000001440
13	11	0.100	0.0206	0.0100	0.0021	0.00000005760
14	$<h\_i$	0.600				
15	$<<t>\_i{\char`\^}2>=$		0.1221			
16			Sum of $h\_i{\char`\^}2=$		5.0600	
17			Sum of $h\_i{*}<t>\_i{\char`\^}2=$		1.0304	
18					Sum of $d\_i{\char`\^}2=$	0.0000170
19						
20					Best slope m =	0.20363439
21					Standard error in slope m =	0.00057962
22					Standard error in g =	0.02795580
23					Standard error in g to 1 sf =	0.03
24					$g =2/m =$	9.8215
25						
26	**Answer. Acceleration due to gravity is $g = 9.82 \pm 0.03$ m/s^2**					

Fig. 6.18 *Excel* spreadsheet to find g by first determining the best m and its error using formulae (4.14) to (4.16) for a line through the origin

The answer from the *Excel* spreadsheet in Fig. 6.17, obtained by approach 2, is:

Acceleration due to gravity is $g = 9.82 \pm 0.03$ ms^{-2}

Since the 1st (rounded) significant figure of the standard error in g corresponds to the 3rd significant figure of g, the calculated value of g is given to 3 sf.

We will comment on the apparent extra accuracy using approach 2 in the overall comment section, Sect. 6.2.5, for Example 2.

6.2.2   Maple Program for the Solution of Example 2.
Determination of the Acceleration due to Gravity

The *Maple* program for the solution of Example 2 follows. For introductory aspects on the *Maple* language, see subsection 6.1.2 in which *Maple* is used for Example 1. Additional commands that are needed for Example 2 are explained in the solution program that follows.

[*MAPLE* PROGRAM FOR THE SOLUTION OF
EXAMPLE 2. To determine the acceleration due to
gravity by measuring the time t for an object to fall
through various heights h and by using the
methodof least squares

[APPROACH 1. Determination of the best m and
c and their errors using formulae (4.5), (4.6) and
(4.10) to (4.13)
  [
  [A colon placed at the end of a command line suppresses output. A semicolon at the end of the command line or no punctuation at the end of the
  command line shows output. You can choose a cell to be a text cell by
  pressing the 'T' button in the menu bar. A new input cell is chosen either
  by pressing return after an existing input cell, or by pressing the button
  '[>'.
  [
  [To access plotting, curve fitting and histogram plotting, the packages
  plots, *CurveFitting* and *Statistics* must be loaded with the
  with(package name) command
[> *restart*
  [The *restart* command clears memory. It is always a good idea to start a
  program with this command
[> *with(plots)*:
[> *with(Statistics)*:
[> *with(CurveFitting)*:
  [
  [We will use both arrays and lists, whichever is more convenient for a
  particular calculation.
  [

[The heights in metres are entered as a *Maple* list labelled gheights:

[> *gheightsi* := [1.1, 1.0, 0.9, 0.8, 0.7, 0.6, 0.5, 0.4, 0.3, 0.2, 0.1]:

[> *gheights* := *sort*(*gheights*)

$$gheights := [0.1, 0.2, 0.3, 0.4, 0.5, 0.6, 0.7, 0.8, 0.9, 1.0, 1.1]$$

(1)

[

[Calculation of the mean of the heights contained in the list *gheights*

[> $gheightsmean = \dfrac{sum(gheights[i], \, i=1..11)}{11}$

$$gheightsmean := 0.6000000000$$

(2)

[

[An array can be defined by *times* := *array*(1..4, [1, 4, 5, 6]), where values are assigned at the outset, or it can be defined without assigning values as in *times* := (*array*(1..11). In the latter case, values are assigned in separate commands. Note that the array definition *array*(1..11) can be left out. Assigning values to each element of an array automatically defines the array.

[

[5 times of fall readings were taken for each height. These times are entered as a *Maple* list in the array *times*[*i*]. The *times* array is defined by assigning it values:

[> *times*[11] := [0.475, 0.470, 0.470, 0.471, 0.477]:

[> *times*[10] := [0.456, 0.451, 0.455, 0.449, 0.454]:

[> *times*[9] := [0.431, 0.433, 0.435, 0.427, 0.421]:

[> *times*[8] := [0.395, 0.412, 0.406, 0.397, 0.405]:

[> *times*[7] := [0.376, 0.364, 0.375, 0.382, 0.381]:

[> *times*[6] := [0.351, 0.347, 0.348, 0.352, 0.346]:

[> *times*[5] := [0.320, 0.323, 0.316, 0.318, 0.326];

[> *times*[4] := [0.282, 0.276, 0.285, 0.277, 0.283];

[> *times*[3] := [0.250, 0.255, 0.246, 0.251, 0.247];

[> *times*[2] := [0.202, 0.197, 0.194, 0.209, 0.206];

[> *times*[1] := [0.153, 0.133, 0.149, 0.151, 0.131];

[

[The mean of the 5 times in each 1-dimensional array element *times*[*i*] is calculated by summing the 5 times and dividing by 5. The mean times are placed in the array *timesmean*[*i*] using a **for** loop

[> **for** *j* **from** 1 **to** 11 **do**

$$timesmean[j] := \frac{sum(times[j][i], \, i=1..5)}{5}$$

end do:

[

[The mean times are squared to obtain $< t >_i^2$. These values are entered into the array *timesmeansq* using a *for* loop

[> **for** k **from** 1 **to** 11 **do**
　　$timesmeansq[k] := timesmean[k]^2$
　　end do:

　[
　[To calculate the slope m of the best straight line and where it cuts the x-axis formulae (4.5) and (4.6) are used with $x_i = h_i$ and $y_i = <t>_i^2$, where h_i are the heights contained in the list $gheights$, and where $<t>_i^2$ are the squared times contained in the array $timesmeansq$.

　[
　[Calculation of the mean of the squares of mean times contained in $timesmeansq$:

[> $timesmeansq\_mean := \dfrac{sum(timesmeansq[z], z=1..11)}{11}$:

　[
　[Calculation of $(x_i - <x>).y_i = (h_i - <h>) <t>_i^2$, where $<h> =$ mean of $gheights$
　[$htsq := [(0.1 - gheightsmean) \cdot timesmeansq[1],$
　　$(0.2 - gheightsmean) \cdot timesmeansq[2], (0.3 - gheightsmean) \cdot$
　　$timesmeansq[3],$
　　$(0.4 - gheightsmean) \cdot timesmeansq[4], (0.5 - gheightsmean) \cdot$
　　$timesmeansq[5],$
　　$(0.6 - gheightsmean) \cdot timesmeansq[6], (0.7 - gheightsmean) \cdot$
　　$timesmeansq[7],$
　　$(0.8 - gheightsmean) \cdot timesmeansq[8], (0.9 - gheightsmean) \cdot$
　　$timesmeansq[9],$
　　$(1.0 - gheightsmean) \cdot timesmeansq[10], (1.1 - gheightsmean) \cdot$
　　$timesmeansq[11]]$:

　[
　[Calculation of the sum of $(x_i - <x>).y_i = (h_i - <h>) <t>_i^2$:
[> $htsqsum := sum(htsq[e], e = 1..11)$:

　[
　[Calculation of $(x_i - <x>)^2 = (h_i - <h>)^2$:
[> $gheightsminmnsq := [(0.1 - gheightsmean)^2,$
　　$(0.2 - gheightsmean)^2, (0.3 - gheightsmean)^2, (0.4 - gheightsmean)^2,$
　　$(0.5 - gheightsmean)^2, (0.6 - gheightsmean)^2, (0.7 - gheightsmean)^2,$
　　$(0.8 - gheightsmean)^2, (0.9 - gheightsmean)^2, (1.0 - gheightsmean)^2,$
　　$(1.1 - gheightsmean)^2]$:

　[
　[Calculation of $D = sum((h_i - <h>)^2)$:
[> $DD := sum(gheightsminmnsq[m], m = 1..11)$:

　[
　[$htsqsum$ and DD are substituted into formula (4.5) to get the slope m of the best straight line:

[> $best\_slope := \dfrac{htsqsum}{DD}$

$$best\_slope := 0.2039810000 \qquad (3)$$

[

[Calculation of c, where the best line cuts the y-axis: $c =< y > -m < x >=< t >_i^2 -m < h >$

[> $c := timesmeansq\_mean - best\_slope * gheightsmean$

$$c := -0.0002779818 \qquad (4)$$

[The acceleration g due to gravity is given by $g = 2h/t^2 = 2/best\_slope$

[> $g := \dfrac{2}{best\_slope}$

$$g := 9.804834764 \qquad (5)$$

[

[The standard error in the slope is found from formula (4.10), after first finding d_i given by formula (4.13)

[> $d\_i = [timesmeansq[1] - 0.1 \cdot best\_slope - c, timesmeansq[2] - 0.2 \cdot best\_slope - c,$

$timesmeansq[3] - 0.3 \cdot best\_slope - c, timesmeansq[4] - 0.4 \cdot best\_slope - c,$

$timesmeansq[5] - 0.5 \cdot best\_slope - c, timesmeansq[6] - 0.6 \cdot best\_slope - c,$

$timesmeansq[7] - 0.7 \cdot best\_slope - c, timesmeansq[8] - 0.8 \cdot best\_slope - c,$

$timesmeansq[9] - 0.9 \cdot best\_slope - c, timesmeansq[10] - 1.0 \cdot best\_slope - c,$

$timesmeansq[11] - 1.1 \cdot best\_slope - c]$:

[

[Calculation of d_i^2:

[> $d\_i\_sq := [d\_i[1]^2, d\_i[2]^2, d\_i[3]^2, d\_i[4]^2, d\_i[5]^2, d\_i[6]^2, d\_i[7]^2,$
$d\_i[8]^2, d\_i[9]^2, d\_i[10]^2, d\_i[11]^2]$:

[

[Determination of the sum of d_i^2:

[> $d\_i\_sq\_sum := sum(d\_i\_sq[h], h = 1..11)$

[

[The standard error in the slope is given by

[> $St\_Error\_in\_m := \left(\dfrac{1}{11-2} \cdot \dfrac{d\_i\_sq\_sum}{DD} \right)^{0.5}$

$$St\_Error\_in\_m := 0.001299147886 \qquad (6)$$

[

[The standard error in c is

[> $Dc := \left(\left(\dfrac{1}{11} + \dfrac{gheightsmean^2}{DD} \right) \cdot \left(\dfrac{d\_i\_sq\_sum}{11-2} \right) \right)^{0.5}$

$$Dc := 0.0008811249659 \qquad (7)$$

[

[The standard error in g is found from the standard error in the slope m using formula (3.19) for errors in inverse proportionalities

$[>$ $sterrg := \dfrac{g \cdot St\_Error\_in\_m}{best\_slope}$

$$Sterrg := 0.06244665121 \tag{8}$$

[
[**Answer. The acceleration due to gravity is** g **= 9.80 pm 0.06 ms^{-2}**

[Since the 1st (rounded) significant figure of the standard error in g corresponds to the 3rd significant figure of g, the calculated value of g is given to 3 sf.

[
[**APPROACH 2. Determination of the best m and its error using formulae (4.14) to (4.16) for a line through the origin**

[
[Calculation of the sum of products of heights in $gheights$ and the mean times squared in $timesmeansq$, i.e., calculation of $x_i . y_i = h_i . < t >_i^2$:

$[>$ $htsqsum\_origin := sum(gheights[i] \cdot timesmeansq[i], i = 1..11):$

[
[Calculation of the sum of the squares of the heights in $gheights$

$[>$ $gheightssqsum := sum(gheights[i]^2, i = 1..11):$

[
[Calculation of d_i given by formula (4.16):

$[>$ $dd\_i := [timesmeansq[1] - 0.1 \cdot best\_slope, timesmeansq[2] - 0.2 \cdot$
$best\_slope,$
$timesmeansq[3] - 0.3 \cdot best\_slope, timesmeansq[4] - 0.4 \cdot best\_slope,$
$timesmeansq[5] - 0.5 \cdot best\_slope, timesmeansq[6] - 0.6 \cdot best\_slope,$
$timesmeansq[7] - 0.7 \cdot best\_slope, timesmeansq[8] - 0.8 \cdot best\_slope,$
$timesmeansq[9] - 0.9 \cdot best\_slope, timesmeansq[10] - 1.0 \cdot best\_slope,$
$timesmeansq[11] - 1.1 \cdot best\_slope]:$

[

[Calculation of d_i^2:

$[>$ $dd\_i\_sq := [dd\_i[1]^2, dd\_i[2]^2, dd\_i[3]^2, dd\_i[4]^2, dd\_i[5]^2, dd\_i[6]^2,$
$dd\_i[7]^2, dd\_i[8]^2, dd\_i[9]^2, dd\_i[10]^2, dd\_i[11]^2] :$

[
[Determination of the sum of d_i^2:

$[>$ $dd\_i\_sq\_sum := sum(dd\_i\_sq[h], h = 1..11):$

[
[The slope of the best line through the origin is found from formula (4.14):

$$[> \; best\_slope\_origin := \frac{htsqsum\_origin}{gheightssqsum}$$

$$best\_slope\_origin := 0.2036184150 \qquad\qquad (9)$$

[
[The acceleration g due to gravity is calculated from the slope of the best line through the origin:

$$[> \; g\_origin := \frac{2}{best\_slope\_origin}$$

$$g\_origin := 9.822294314; \qquad\qquad (10)$$

[
[The standard error in the slope for the line through the origin is given by

$$[> \; St\_Error\_in\_m\_origin := \left(\frac{1}{11-1} \cdot \frac{dd\_i\_sq\_sum}{gheightssqsum} \right)^{0.5}$$

$$St\_Error\_in\_m\_origin := 0.0005890819311 \qquad\qquad (11)$$

[
[The standard error in g is found from the standard error in the slope m using formula (3.19) for errors in inverse proportionalities

$$[> \; sterrorgorigin := \frac{g\_origin \cdot St\_Error\_in\_m\_origin}{best\_slope\_origin}$$

$$sterrorgorigin := 0.02841656587 \qquad\qquad (12)$$

[
[**Answer. The acceleration due to gravity is g = 9.82 pm 0.03 ms^{-2}**

[Since the 1st (rounded) significant figure of the standard error in g corresponds to the 3rd significant figure of g, the calculated value of g is given to 3 sf.

[
[We will comment on the extra accuracy of approach 2 in the comments section, §6.2.5

[
[# The Graphs
[
[## Plot of Data Points
[The following is a plot of the squares of the mean times contained in the array $timesmeansq[i]$ versus the heights contained in the list $gheights$.
[First, the array $timesmeansq$ is converted into a list

$[> \; timesmeansqList := [timesmeansq[1], \; timesmeansq[2], \; timesmeansq[3],$
$\quad timesmeansq[4], \; timesmeansq[5], \; timesmeansq[6], \; timesmeansq[7],$
$\quad timesmeansq[8], \; timesmeansq[9], \; timesmeansq[10], \; timesmeansq[11]];$

[
$[> \; PG1 := pointplot(gheights, \; timesmeansqList,$
$\quad axesfont = [Calibri, \; roman, \; 12], titlefont = [Calibri, \; roman, \; 18],$
$\quad labelfont = [Calibri, \; roman, \; 14], \; title = \text{``Time Squared versus}$

Height",
$axes = frame$, $labels = $ ["Height (m)", "Time squared (seconds sq.)"],
$labeldirections = [horizontal, vertical]$, $color = $ "Orange",
$symbolsize = 20$, $legendstyle = [location = right]$, $legend = $ "Data
points"):
[The plot is suppressed because it will be shown below combined with
other plots.

[
[## The Graph for Approach 1
[The following defines the line with the best slope given by $best\_slope$,
which cuts the y- axis at $c = $ -0.00028:
[> $TSQ := (h) \rightarrow best\_slope \cdot h + c$:

$$TSQ := h \rightarrow best\_slope\ h + c \tag{13}$$

[
[Plot of the line TSQ:
[> $PG2 := plot(TSQ(h), h = 0..1.2, axesfont = [Calibri, roman, 12]$,
$titlefont = [Calibri, roman, 18]$, $labelfont = [Calibri, roman, 14]$,
$title = $ "Time Squared versus Height", $axes = frame$,
$labels = $ ["Height (m)", "Time squared (seconds sq.)"],
$labeldirections = [horizontal, vertical]$, $color = $ "Red",
$legendstyle = [location = right]$, $legend = $ "Best Line");
[The plot is suppressed because it will be shown below combined with
other plots.
[Combined plot of TSQ and the data points
[> $display(PG1, PG2)$:

[
[## The Graph for Approach 2
[Since the line passes through the origin the equation of the best line has
slope given by $best\_slope\_origin$ and $c = 0$;
[> $TSQorigin := (h) \rightarrow best\_slope\_origin \cdot h$:

[
[Plot of the line $TSQorigin$ through the origin
[> $PG3 := plot(TSQorigin(h), h = 0..1.2, axesfont = [Calibri,$
$roman, 12]$,
$titlefont = [Calibri, roman, 18]$, $labelfont = [Calibri, roman, 14]$,
$title = $ "Time Squared versus Height", $axes = frame$,
$labels = $ ["Height (m)", "Time squared (seconds sq.)"],
$labeldirections = [horizontal, vertical]$, $color = $ "Blue",
$legendstyle = [location = right]$, $legend = $ "Best Line through the
origin",
$linestyle = [dot]$):
Combined plot of TSQ, $TSQorigin$
and the data points

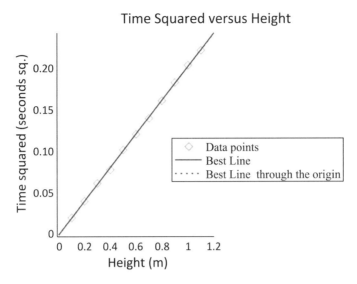

Fig. 6.19 *Maple* plot of the line with the best *m* and best *c*, the line with the best *m* passing through the origin and the data points

The plot is shown in Fig. 6.19
[> *display*(*PG*1, *PG*2, *PG*3)

6.2.3   Mathematica Program for the Solution of Example 2.
Determination of the Acceleration due to Gravity

The *Mathematica* program for the solution of Example 2 follows. For introductory aspects on the *Mathematica* language, see subsection 6.1.3 in which *Mathematica* is used for Example 1. Additional commands that are needed for Example 2 are explained in the solution program that follows.

MATHEMATICA **PROGRAM FOR THE SOLUTION OF EXAMPLE 2. To determine the acceleration due to gravity by measuring the time *t* for an object to fall through various heights *h* and by using the method of least squares**

APPROACH 1. Determination of the best m and c and their errors using formulae (4.5), (4.6) and (4.10) to (4.13)

Note: Terminating a command line with a semicolon suppresses output, while leaving a command line open shows output. Enclosing a command line with (*... *) suppresses execution of a command.

In[1]:= `clear;`

The heights in metres are entered as a *Mathematica* list called *gheightsi*

In[2]:= `gheightsi ={1.1,  1.0,  0.9,  0.8,  0.7,  0.6,  0.5,`
`0.4,  0.3,  0.2,  0.1};`

In[3]:= `gheights = Sort[gheightsi]`

Out[3]:= `{0.1,  0.2,  0.3,  0.4,  0.5,  0.6,  0.7,  0.8,  0.9,  1.,`
`1.1}`

The following extracts the fourth element of the list *gheights*:

In[4]:= `gheights[[4]]`

Out[4]:= `0.4`

Calculation of the mean of the heights contained in the list *gheights*

In[5]:= `gheightsmean = Sum[gheights[[i]], {i, 1, 11}]/11;`

5 times of fall were taken for each height and entered as a *Mathematica* list. These lists are then entered into the array *times*

Creation of the array *times*

In[6]:= `Array[times, 11];`

Assignment of values to the array elements of the array *times*

In[7]:= `times[1] = {0.153, 0.133, 0.149, 0.151, 0.131};`

In[8]:= `times[2] = {0.202, 0.197, 0.194, 0.209, 0.206};`

In[9]:= `times[3] = {0.250, 0.255, 0.246, 0.251, 0.247};`

In[10]:= `times[4] = {0.282, 0.276, 0.285, 0.277, 0.283};`

In[11]:= `times[5] = {0.320, 0.323, 0.316, 0.318, 0.326};`

In[12]:= `times[6] = {0.351, 0.347, 0.348, 0.352, 0.346};`

In[13]:= `times[7] = {0.376, 0.364, 0.375, 0.382, 0.381};`

In[14]:= `times[8] = {0.395, 0.412, 0.406, 0.397, 0.405};`

In[15]:= `times[9] = {0.431, 0.433, 0.435, 0.427, 0.421};`

In[16]:= `times[10] = {0.456, 0.451, 0.455, 0.449, 0.454};`

In[17]:= `times[11] = {0.475, 0.470, 0.470, 0.471, 0.477};`

The following extracts the 8th element of the array *times*. Notice that each element of the array times is a list. Also notice the difference from extracting elements from a list, as in the example *gheights[[i]]* above,

In[18]:= `times[8];`

Out[18]:= `{0.395, 0.412, 0.406, 0.397, 0.405}`

The following extracts the 3rd element of the list *times[8]*

In[19]:= `times[8][[3]]`

Out[19]:= `0.406`

The mean of the 5 times in each 1-dimensional array *times[i]* is calculated by summing them and dividing by 5. The mean time for each
height is placed in the array *timesmean* using a *For* loop

In[20]:= `For[i = 1, i < 12, i++, timesmean[i] =`
`Sum[ times[i][[j]], {j, 1, 5} ]/5];`

Notice the syntax of the *For* loop. The i counts the number of execu-
tions of the loop. The $i < 12$ is a test which terminates the loop once
$i = 12$, The $i + +$ increments the counter i after each loop execution.
Any letter not used by *Mathematica* can be used as the counter. The
timesmean[i] = *times[i][[j]]* is the expression to be executed with each
pass of the loop.

The contents of the array *timesmean* can be shown using a *Print* command and a *For* loop The syntax is shown in the following command.
Output is suppressed.

In[22]:= `For[i = 1, i < 12, i++, Print [timesmean[i]]];`

The mean times are squared to obtain $< t >\_i\char`\^2$. These values are
entered into the array *timesmeansq* using a *For* loop.

In[23]:= `For [i = 1, i < 12, i++, timesmeansq[i] = timesmean`
`[i]^2];`

To calculate the slope m of the best straight line and where it cuts the
x-axis, formulae (4.5) and (4.6) are used with $x\_i = h\_i$ and $y\_i = < t >\_i\char`\^2$, where h_i are the heights contained in the list *gheights*, and
where $< t >\_i\char`\^2$ are the mean squared times in the array *timeseansq*

Calculation of the mean of the squares of mean times contained in
timesmeansq

In[24]:= `timesmeansqmean = Sum[timesmeansq[i], {i, 1, 11}]`
`/11;`

Calculation of $(x\_i- < x >)y\_i = (h\_i- < h >) < t >\_i\char`\^2$, where $< h > =$ mean of *gheights*

In[25]:= `htsq = {(0.1 - gheightsmean)*timesmeansq[1],`
`(0.2 - gheightsmean)*timesmeansq[2],`
`(0.3 - gheightsmean)*timesmeansq[3],`
`(0.4 - gheightsmean)*timesmeansq[4],`
`(0.5 - gheightsmean)*timesmeansq[5],`
`(0.6 - gheightsmean)*timesmeansq[6],`
`(0.7 - gheightsmean)*timesmeansq[7],`
`(0.8 - gheightsmean)*timesmeansq[8],`
`(0.9 - gheightsmean)*timesmeansq[9],`
`(1.0 - gheightsmean)*timesmeansq[10],`
`(1.1 - gheightsmean)*timesmeansq[11]};`

Calculation of the sum of $(x\_i- < x >)y\_i = (h\_i- < h >) < t >\_i\char`\^2$

In[26]:= `htsqsum = Sum[htsq[[i]], {i, 1, 11}];`

Calculation of $(x\_i- <x>)^2 = (h\_i- <h>)^2$

In[27]:= gheightsminmnsq = {(0.1 - gheightsmean)^2, (0.2 -
gheightsmean)^2,
(0.3 - gheightsmean)^2, (0.4 - gheightsmean)^2,
(0.5 - gheightsmean)^2, (0.6 - gheightsmean)^2,
(0.7 - gheightsmean)^2, (0.8 - gheightsmean)^2,
(0.9 - gheightsmean)^2, (1.0 - gheightsmean)^2,
(1.1 - gheightsmean)^2};

Calculation of $D = sum((h\_i- <h>)^2)$

In[28]:= DD = Sum[gheightsminmnsq[[i]], {i, 1, 11}];

htsqsum and *DD* are substituted into formula (4.5) to get the slope *m*
of the best straight line

In[29]:= bestslope = htsqsum/DD

Out[29]:=   0.203981

Calculation of *c*, where the best line cuts the *y*-axis:

$c = <y> -m<x> = <<t>\_i^2> -m<h>$

In[30]:= c = timesmeansqmean - bestslope*gheightsmean

Out[30]:=   -0.000277982

The acceleration *g* due to gravity is given by $g = 2h/t^2 = 2/bestslope$

In[31]:= g = 2/bestslope

Out[31]:=   9.80483

The standard error in the slope is found from formula (4.10), after first
finding d_i given by formula (4.13)

In[32]:= di = { timesmeansq[1] - bestslope*0.1 - c,
timesmeansq[2] - bestslope*0.2 - c,
timesmeansq[3] - bestslope*0.3 - c,
timesmeansq[4] - bestslope*0.4 - c,
timesmeansq[5] - bestslope*0.5 - c,
timesmeansq[6] - bestslope*0.6 - c,
timesmeansq[7] - bestslope*0.7 - c,
timesmeansq[8] - bestslope*0.8 - c,
timesmeansq[9] - bestslope*0.9 - c,
timesmeansq[10] - bestslope*1.0 - c,
timesmeansq[11] - bestslope*1.1 - c};

Calculation of d_i^2

In[33]:= disq = d_i^2 ;

Determination of the sum of d_i^2

In[34]:= disqsum = Sum[disq[[i]], {i, 1, 11}];

The standard error in the slope *m* is given by

In[35]:= StErrm = (1/(11 - 2) *(disqsum/DD))$^{0.5}$

Out[35]:=   0.00129915

The standard error in *c* is found from Eq. (4.11)

In[36]:= Dc = ((1/11 + gheightsmean^2/DD)*(disqsum/
(11 - 2)))^0.5 ;

Out[36]:=   `0.000881125`

The standard error in g is found from the standard error in the slope m using formula (3.19) for errors in inverse proportionalities

In[37]:= `sterrg = g*StErrm/bestslope`

Out[37]:=   `0.0624467`

Answer. The acceleration due to gravity is g = 9.80 pm 0.06 m/s^2

Since the 1st (rounded) significant figure of the standard error in g corresponds to the 3rd significant figure of g, the calculated value of g is given to 3 sf.

APPROACH 2. Determination of the best m and its error using formulae (4.14) to (4.16) for lines through the origin

Calculation of the sum of products of heights in *gheights* and the mean times squared in *timesmeansq*, i.e., calculation of $x\_i \cdot y\_i = h_i < t > \_i$^2

In[38]:= `htsqorigin = {0.1*timesmeansq[1], 0.2*timesmeansq`
`[2], 0.3*timesmeansq[3], 0.4*timesmeansq[4],`
`0.5*timesmeansq[5], 0.6*timesmeansq[6],`
`0.7*timesmeansq[7], 0.8*timesmeansq[8],`
`0.9*timesmeansq[9], 1.0*timesmeansq[10],`
`1.1*timesmeansq[11]};`

Calculation of the sum of $h\_i < t >\_i$^2

In[39]:= `htsqsumorigin = Sum[htsqorigin[[i]], {i, 1, 11}];`

Calculation of the h_i^2 corresponding to the x_i^2 in formula (4.14)

In[40]:= `gheightssq = gheights^2;`

Calculation of the sum of the squares of the heights in *gheights*

In[41]:= `gheightssqsum = Sum[gheightssq[[i]], {i, 1, 11}];`

Calculation of d_i given by formula (4.16)

In[42]:= `ddi = { timesmeansq[1] - bestslope*0.1,`
`timesmeansq[2] - bestslope*0.2, timesmeansq[3] -`
`bestslope*0.3,`
`timesmeansq[4] - bestslope*0.4, timesmeansq[5] -`
`bestslope*0.5,`
`timesmeansq[6] - bestslope*0.6, timesmeansq[7] -`
`bestslope*0.7,`
`timesmeansq[8] - bestslope*0.8, timesmeansq[9] -`
`bestslope*0.9,`

```
        timesmeansq[10] - bestslope*1.0, timesmeansq[11] -
        bestslope*1.1};
```
Calculation of d_i^2

In[43]:= `ddisq = ddi^2;`

Determination of the sum of d_i^2

In[44]:= `ddisqsum = Sum[ddisq[[i]], {i, 1, 11}];`

The slope of the best line through the origin is found from formula (4.14)

In[45]:= `bestslopeorigin = htsqsumorigin/gheightssqsum;`

Out[45]:= 0.203618

The acceleration g due to gravity is calculated from the slope of the best line through the origin

In[46]:= `gorigin = 2/bestslopeorigin;`

Out[46]:=   9.82229

The standard error in the slope for a line through the origin is found from formula (4.15)

In[47]:= `StErrmorigin = ((1/(11 - 1))*ddisqsum/`
         `gheightssqsum )^0.5;`

Out[47]:=   0.000589082

The standard error in g is found from the standard error in the slope m using formula (3.19) for errors in inverse proportionalities

In[48]:= `sterrgorigin = g*StErrmorigin/bestslopeorigin`

Out[48]:=   0.0283661

Answer. The acceleration due to gravity is $g =$ 9.82 pm 0.03 m/s^2

Since the 1st (rounded) significant figure of the standard error in g corresponds to the 3rd significant figure of g, the calculated value of g is given to 3 sf.

We will comment on the extra accuracy of approach 2 in subsection, §6.2.5

The Graphs

Plot of Data Points

The following is a plot of the data points, i.e., the squares of the mean times for each height contained in the array *timesmeansq* versus the heights contained in the list *gheights*.

In[49]:= `PG1 = ListPlot[ { {0, 0}, {0.1, 0.0206}, {0.2,`
         `0.0406},`
         `{0.3, 0.0624},{0.4, 0.0787}, {0.5, 0.1028}, {0.6,`
         `0.1217},`
         `{0.7, 0.1411}, {0.8, 0.1624}, {0.9, 0.1844}, {1.0,`

```
0.2052},
{1.1, 0.2234} }, PlotLegends → Placed[{"Data
points"}, Right],
PlotStyle → {Black, PointSize[0.03]}, AspectRatio
→ 5/4,
LabelStyle → {FontFamily → "Calibri", 14,
GrayLevel[0]},
AxesStyle → Thick, TicksStyle → Directive[Thin]];
```

The Graph for Approach 1

The following defines the line with the best slope given by *bestslope*, which cuts the *y*- axis at $c = -0.00028$:

```
In[50]:= TSQ[h_] := bestslope*h + c;
```
A Plot of the line TSQ
```
In[51]:= PG2 = Plot[TSQ[h], {h, 0, 1.1}, AxesLabel →
HoldForm["Height (m)"],
HoldForm["Times squared (seconds sq.)"],
PlotLabel → HoldForm["Time Squared versus
Height"],
BaseStyle → {FontFamily → "Calibri", FontSize
→ 16},
LabelStyle → {FontFamily → "Calibri", 14,
GrayLevel[0]},
PlotStyle → {Blue, Thick},
PlotLegends → Placed[{"Best line"}, Right],
AspectRatio → 5/4
AxesStyle → Thick, TicksStyle → Directive[Thin]];
```

The Graph for Approach 2

Since the line passes through the origin the equation of the line with best slope *bestslopeorigin* is given by

```
In[52]:= TSQorigin = bestslopeorigin*h;
```

A Plot of the line TSQorigin
```
In[53]:= PG3 = Plot[TSQorigin, {h, 0, 1.1},
PlotStyle → {Orange, Dashing[.05], Thick},
AxesLabel → HoldForm["Height (m)"],
HoldForm["Times squared (seconds sq.)"],
PlotLabel → HoldForm["Time Squared versus
Height"],
LabelStyle → {FontFamily → "Calibri", 14,
GrayLevel[0]},
PlotLegends → Placed[{"Line through the origin"},
```

Fig. 6.20 *Mathematica* plot of the line with the best *m* and best *c*, the line with the best *m* passing through the origin and the data points

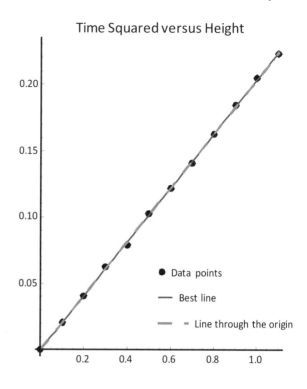

```
                                                  Right],
                                                  AspectRatio → 5/4, AxesStyle → Thick,
                                                  TicksStyle → Directive[Thin]];
```
A combined plot of *TSQ*, *TSQorigin* and the data points

In[54]:= `Show[PG1, PG2, PG3, PlotLabel → HoldForm["Time`
`Squared versus Height"], BaseStyle → {FontFamily`
`→ "Calibri", FontSize → 16} ]`

We see that the difference between the best line and the best line through the origin is too small to be visible graphically. As a result the error indicated by the distance the best line misses the origin is also too small to be seen visually. Note that the axes labels in the original plots do not appear in the combined plot.

The plot is shown in Fig. 6.20

6.2.4   *Matlab Program for the Solution of Example 2.*
Determination of the Acceleration due to Gravity

The *Matlab* program for the solution of Example 2 follows. For introductory aspects on the *Matlab* language, see subsection 6.1.4 in which *Matlab* is used for Example 1. Additional commands that are needed for Example 2 are explained in the solution program that follows

```
% Matlab Program for the Solution of Example 2. To determine the acceleration due
% to gravity by measuring the time t for an object to fall through various heights h
% and by using the method of least squares.

% SOME ASPECTS OF THE MATLAB LANGUAGE FOLLOW:

% The following command sorts the array 'times':
% timessorted = sort(times)

% The following command sums the array 'times':
% timessum = sum(times)

% As explained in example 1, Matlab deals with arrays so we will prefer to refer to
% arrays of data rather than lists of data.

% Recall that the dot versions of the usual arithmetic operators, i.e., .* and ./ and .^
% apply to each element of an array.

% xxxxxxxxxxxxxxxxxxxxxxxxxxxxxxxxxxxxxxxxxxxxxxxxxxxxxxxxxxxxxxxxxxxxxxxx
% APPROACH 1. Determination of the best m and c and their errors using formulae
% (4.5), (4.6) and (4.10) to (4.13)
% xxxxxxxxxxxxxxxxxxxxxxxxxxxxxxxxxxxxxxxxxxxxxxxxxxxxxxxxxxxxxxxxxxxxxxxx
clear
format long

% The heights in metres are entered as a Matlab array labelled 'gheights':
gheights = [0.1:0.1:1.1]
% The colon operator in the above command line generates a list of numbers from
% 0.1 to 1.1 in steps of 0.1. The syntax for generating a list with a colon operator is:
% [initial value, step size, final value].

% The mean of 'gheights' is found using Matlabs 'sum' command, then entered into
% the array gheightsmean
gheightsmean = sum(gheights)/11

% The 11 sets of five times for each height are entered into 11 arrays labelled 'times'
```

times1 = [0.153, 0.133, .149, 0.151, 0.131]
times2 = [0.202, 0.197, 0.194, 0.209, 0.206]
times3 = [0.250, 0.255, 0.246, 0.251, 0.247]
times4 = [0.282, 0.276, 0.285, 0.277, 0.283]
times5 = [0.320, 0.323, 0.316, 0.318, 0.326]
times6 = [0.351, 0.347, 0.348, 0.352, 0.346]
times7 = [0.376, 0.364, 0.375, 0.382, 0.381]
times8 = [0.395, 0.412, 0.406, 0.397, 0.405]
times9 = [0.431, 0.433, 0.435, 0.427, 0.421]
times10 = [0.456, 0.451, 0.455, 0.449, 0.454]
times11 = [0.475, 0.470, 0.470, 0.471, 0.477]

% Calculation of the mean of the 5 times in each 1-dimensional array 'times i' is
% done by summing each and dividing by 5. The mean time for each height is placed
% in the array 'timesmean'
timesmean = [sum(times1), sum(times2), sum(times3), sum(times4), ...
sum(times5),sum(times6),sum(times7),sum(times8), sum(times9), ...
sum(times10),sum(times11)]./5
% Note: The ellipsis indicate that a command is continued

% The mean times are squared to obtain $< t >\_i^2$. These values are entered into
% the array 'timesmeansq'
timesmeansq = timesmean. ^2
% To calculate the slope m of the best straight line and where it cuts the x-axis
% formulae (4.5) and (4.6) are used with $x\_i = h\_i$ and $y\_i = < t >\_i^2$, where
% h_i are the heights contained in the list 'gheights', and where $< t > \_i^2$ are the
% squared mean times in the array 'timeseansq'

% Calculation of the mean of the squares of the mean times contained in
% 'timesmeansq'
timesmeansqmean = sum(timesmeansq)/11

% Calculation of $(x\_i - < x >)y\_i = (h\_i - < h >) < t > \_i^2$,
% where $< h > =$ mean of 'gheights'
htsq = (gheights-gheightsmean).*timesmeansq

% The ability to operate element by element on arrays and matrices by using the dot
% versions of arithmetic operators, as mentioned earlier, is one of the big advantages
% of Matlab. Thus, using .* in the above command line we were able to first subtract
% 'gheightsmean' from each element of the 'gheights' array then multiply each result
% by 'timesmeansq', making the calculation much easier.

% Calculation of the sum of $(x\_i - < x >)y\_i = (h\_i - < h >)< t >\_i^2$
htsqsum = sum(htsq)

Calculation of $(x\_i- <x>)^2 = (h\_i- <h>)^2$
gheightsminmnsq = (gheights-gheightsmean).^2

% Calculation of $D = sum((h\_i- <h>)^2)$
DD = sum(gheightsminmnsq)

% 'htsqsum' and 'DD' are substituted into formula (4.5) to get the slope m of the
% best straight line
bestslope = htsqsum/DD
% Calculation of c, where the best line cuts the y-axis, using formula (4.6)
% $c = <y> -m<x> = <t>\_i^2 - m<h>$
c = timesmeansqmean - bestslope*gheightsmean

% The acceleration g due to gravity is given by $g = 2h/t^2 = 2/bestslope$
g = 2/bestslope

% The standard error in the slope is found from formula (4.10), after first finding
% d_i^2 using formula (4.13)
disq = (timesmeansq-gheights.*bestslope-c).^2

% Determination of the sum of d_i^2
disqsum = sum(disq)

% The standard error in the slope is given by
StErrm = (1/(11 - 2).*(disqsum/DD))^0.5

% The standard error in c is
Dc = ((1/11 + gheightsmean.^2/DD).*(disqsum/(11 - 2))).^0.5

% The standard error in g is found from the standard error in the slope m using
% formula (3.19) for errors in inverse proportionalities
Sterrg = g*StErrm/bestslope

% xxxxxxxxxxxxxxxxxxxxxxxxxxxxxxx ANSWER xxxxxxxxxxxxxxxxxxxxxxxxxxxxxxx
% Answer. The acceleration due to gravity is $g = 9.80$ pm 0.06 m/s^2
% xxxxxxxxxxxxxxxxxxxxxxxxxxxxxxxxxxxxxxxxxxxxxxxxxxxxxxxxxxxxxxxxxxxxx
% Since the 1st (rounded) significant figure of the standard error in g corresponds to
% the 3rd significant figure of g, the calculated value of g is given to 3 sf.

% xxxxxxxxxxxxxxxxxxxxxxxxxxxxxxxxxxxxxxxxxxxxxxxxxxxxxxxxxxxxxxxxxxxxx
% APPROACH 2. Determination of the best m and its error using
% formulae (4.14) to (4.16) for lines through the origin.
% xxxxxxxxxxxxxxxxxxxxxxxxxxxxxxxxxxxxxxxxxxxxxxxxxxxxxxxxxxxxxxxxxxxxx

% Calculation of the sum of products of heights in 'gheights' and the mean times

% squared in 'timesmeansq', i.e., calculation of x_i · y_i = h_i < t >_i^2
```
htsqorigin=gheights.*timesmeansq
```

% Calculation of the sum of h_i < t >_i^2
```
htsqoriginsum=sum(htsqorigin)
```

%Calculation of the heights squared corresponding to the x_i^2 in formula (4.14)
```
gheightssq = gheights. ^2
```

% Calculation of the sum of the squares of the heights in 'gheights'
```
gheightssqsum = sum(gheightssq)
```

% Calculation of d_i^2 needed by formula (4.15)
```
ddisq =(timesmeansq-gheights.*bestslope).^2
```

% Determination of the sum of d_i^2
```
ddisqsum = sum(ddisq)
```

% Calculation of the best slope through the origin
```
bestslopeorigin= htsqoriginsum/gheightssqsum
```

%The acceleration g due to gravity is calculated from the slope of the line through
% the origin
```
gorigin = 2/bestslopeorigin
```

%The standard error in the slope for a line through the origin is found from
% formula (4.15)
```
StErrorigin = sqrt( (1/(11-1))*ddisqsum/gheightssqsum )
```

% The standard error in g is found from the standard error in the slope m using
% formula (3.19) for errors in inverse proportionalities
```
Sterrgorigin = gorigin*StErrorigin/bestslopeorigin
```

% xxxxxxxxxxxxxxxxxxxxxxxx ANSWER xxxxxxxxxxxxxxxxxxxxxxxxxxxxxxxxxxxxx
% Answer. The acceleration due to gravity is g = 9.82 pm 0.03 m/s^2
% xxxxxxxxxxxxxxxxxxxxxxxxxxxxxxxxxxxxxxxxxxxxxxxxxxxxxxxxxxxxxxxxxxxxxx
% Since the 1st (rounded) significant figure of the standard error in g corresponds to
% the 3 significant figure of g, the calculated value of g is given to 3 sf.

% xxxxxxxxxxxxxxxxxxxxxxxxxxxxxxxxxxxxxxxxxxxxxxxxxxxxxxxxxxxxxxxxxxxxxx
% xxxxxxxxxxxxxxxxxxxxxxxxxxxxTHE GRAPHSxxxxxxxxxxxxxxxxxxxxxxxxxxxx
% xxxxxxxxxxxxxxxxxxxxxxxxxxxxxxxxxxxxxxxxxxxxxxxxxxxxxxxxxxxxxxxxxxxxxx

% First we recall the syntax of various options in the plot command

% In the plot command 'plot(x,y)', x is a vector of x-axis values to be plotted, and
% y is a vector of y-axis values to be plotted. The x and y vectors must have the
% same dimension. Note that y may also be a function into which the values in the
% x-vector are fed.

% The plot command p(x,y,r) allows various line colours and line types to be chosen
% through choices of r, where r is a character string of options. Two examples follow:
% E.g. 1. plot(x,y,'rd:')plots a red dotted line with a diamond at each data point.
% E.g.2 plot(x,y, 'bo–') plots a dashed blue line with a circle at each data point.

% plot(x1,x2,r1, x2,y2,r2,...,xn,yn,rn) combines the plots (x1,y1) to (xn,yn) with
% colours and line types r1 to rn.

%xxxxxxxxxxxxxxxxxxxxxxxxxxxxxxxxxxxxxxxxxxxxxxxxxxxxxxxxxxxxxxxxxxx
%xxxxxxxx PLOT OF BEST LINE AND DATA POINTS xxxxxxxxxxxxxxxxxxxxx
%xxxxxxxxxxxxxxxxxxxxxxxxxxxxxxxxxxxxxxxxxxxxxxxxxxxxxxxxxxxxxxxxxxx
% The line with best slope m = bestslope = 0.203981, which cuts the y-axis at
% c = -0.000277982, is plotted together with data points (h_i, < t > _i^2)
% contained in the arrays 'gheights' and 'timesmeansq'.

% To do this, the equation TSQmfn(h) of the best straight line is first defined as a
% separate Matlab function file, then saved in an .m file named TSQbestmfn.m.

h = 0:0.1:1.2 % x-vector for the best line plot
fb = TSQbestmfn(h) % Equation of the best straight line
fo = TSQoriginmfn(h) % Equation of the best straight line through the origin.

x1 = gheights % list of heights forming the x-axis values
y1 = timesmeansq % list of the mean times squared forming the y-axis values

% Plot of the best straight line and (h_i, < t > _i^2) data points

subplot(2,2,1)
plot (h,fb,'b',x1,y1,'ro'), title ('Time Squared versus Height'),
xlabel ('Height (m)'), ylabel ('Time squared (s^2)'), grid on,
legend('Best line', 'Data points'),axis([0,1.2,0,0.5])

% subplot(m.n.p) is used to produce m rows and n columns of plots, while p specifies
% the position of the plot. The option 'b' indicates the colour blue. Take note of the
% format for adding a title, xlabel, ylabel, grid and legend.

% xxxxxxxxxxxxxxxxxxxxxxxxxxxxxxxxxxxxxxxxxxxxxxxxxxxxxxxxxxxxxxxxxxx
% xxxxxxxxxxxxxx PLOT OF THE BEST LINE THROUGH THE ORIGIN xxxxxxxx
% xxxxxxxxxxxxxxxxxxxxxxx AND DATA POINTS xxxxxxxxxxxxxxxxxxxxxxxxxxx
% xxxxxxxxxxxxxxxxxxxxxxxxxxxxxxxxxxxxxxxxxxxxxxxxxxxxxxxxxxxxxxxxxxx

% The equation for the best line through the origin has m = 0.203618 and c = 0. Using
% this m and c the equation of the best straight line through the origin is defined in a
% Matlab .m function file, then saved in an .m file named TSQoriginmfn.m.

% Plot of the best straight line through the origin and (h_i, < t > _i^2) data points

subplot(2,2,2)
plot (h,fo,'r',x1,y1,'o'), title ('Time Squared versus Height'),
xlabel ('Height (m)'), ylabel('Time squared (s^2)'), grid on,
legend ('Line through the origin', 'Data points'), axis ([0,1.2,0,0.5])

% xxxxxxxxxxxxxxxxxxxxxxxxxxxxxxxxxxxxxxxxxxxxxxxxxxxxxxxxxxxxxxxxxxxxxxxx
% xxxxxxxxxxxxxxxxxx COMBINED PLOTS xxxxxxxxxxxxxxxxxxxxxxxxxxxxxxxxxx
% xxxxxxxxxxxxxxxxxxxxxxxxxxxxxxxxxxxxxxxxxxxxxxxxxxxxxxxxxxxxxxxxxxxxxxxx
% There are two main ways of combing plots. The first uses the 'hold on' command.
% The 'hold on' command keeps the current plot and adds to it all subsequent plots
% until a 'hold off' or 'hold' (hold toggles between on/off states) command is issued.
% The second, and preferred way, is to use the 'plot' command with multiple
% arguments.

% Combined plot of the best line and the best line through the origin

subplot(2,2,3)
plot(h,fb,'b',h,fo,'y–'), title('Time Squared versus Height'),
xlabel('Height (m)'), ylabel('Time squared (s^2)'), grid on,
legend('Best line', 'Line through the origin'), axis([0,1.2,0,0.5])

% Combined plot of the best line, the best line through the origin and (h_i, < t >_i^2)
% data points

subplot (2,2,4)
plot(h,fb,'b',h,fo,'y–',x1,y1,'ko'), title('Time Squared versus Height'),
xlabel('Height (m)'), ylabel('Time squared (s^2)'), grid on,
legend('Best line','Line through the origin','Data points'),
axis ([0,1.2,0,0.5])

% The plot combination is shown in Fig. 6.21

% xxxxxxxxxxxxxxxxxxxxxxxxxxxxxx End of Program xxxxxxxxxxxxxxxxxxxxxxxxxxxxxxxx

The *Matlab*.m function files TSQbestmfn and TSQoriginmfn used in the above program are defined as follows:

[−] function [TSQb6] = TSQbestmfn(h)

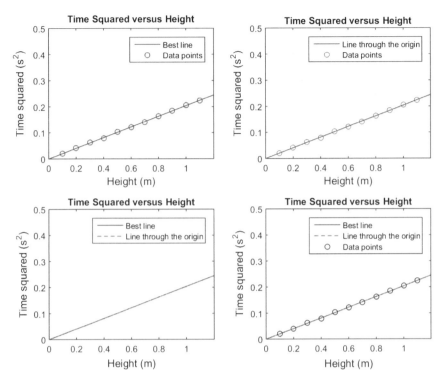

Fig. 6.21 Matlab plots. Top left: Plot of the line with the best *m* and best *c* together with the data points. Top right: Plot of the line with the best *m* passing through the origin together with the data points. Bottom left: Plot of both lines. As can be seen, the difference between them is too small to be seen graphically. Bottom right: Combination of both lines and the data points

```
| % Equation of the best straight line
| TSQb6=0.203981.*h-0.000277982
Lend
```

```
[−] function [TSQo6] = TSQoriginmfn( h )
| % Equation of the best straight line through the origin
| TSQo6=0.203618.*h
Lend
```

6.2.5 Comments on the Solution Programs of Example 2

For approach 1, the answer from all four programs is:

Answer. The acceleration due to gravity is $g = 9.80 \pm 0.06$ ms^{-2}

For approach 2, the answer from all four programs is:

Answer. The acceleration due to gravity is $g = 9.82 \pm 0.03 \ \text{ms}^{-2}$

We see that the standard error in g for the line through the origin is half that for the line with the best m and through the best c. This is because the theoretical input, namely, that the line passes through the origin, is not subject to error, hence improving the error estimate.

The spread of the times is fairly small, indicating reasonable precision. Since we estimate that systematic error to be fairly small, we may conclude that the result is reasonably accurate.

Appendix A
Bibliography

1. Apollonius. *Plane Loci*, (3rd century BC).
2. Aristotle. *Peri ta zōa historiai* (*History of animals*), *Physikē* (*Physics*), *Peri psychēs* (*On the Soul*), *Ēthika Nikomacheia* (*Nicomachean Ethics*) and *Politika* (*Politics*).
3. Arnauld, A. et al. (1560–1619). Port Royal *logic*.
4. Baird, D. C. (1962). *Experimentation: An introduction to measurement theory and experiment design*. New Jersey: Prentice-Hall.
5. Barford, N. C. (1967). *Experimental measurements: Precision, error and truth*. London: Addison-Wesley.
6. Bayes, T. (1763). An essay towards solving a problem in the doctrine of chances. *Philosophical Transactions of the Royal Society of London, 53*, 370–418.
7. Beisbart, C., & Hartmann, S. (2011). *Probabilities in physics*. Oxford: Oxford University Press.
8. Bernoulli, D. (1738). *Hydrodynamica*. Basel.
9. Bernoulli, J. (1713). *Ars conjectandi* (*The art of conjecturing*). Basel.
10. Cardano, J. (1663). *Opera omnia* (10 Vols.). Amsterdam.
11. Carnap, R. (1950). *Logical foundations of probability*. Chicago.
12. Copernicus, N. (1543). *De revolutionibus orbium coelestium libri vi* (*Six books concerning the revolutions of the heavenly orbs*).
13. Descarte, R. (1637). *Géométrie*.
14. Empiricus, S. (204). *Outlines of pyrrhonism*.
15. de Fermat, P. (1679). *Introduction to loci*.
16. Galilei, G. (1632). *Dialogo: dove ne i congressi di quattro giornate si discorre sopra i due massimi sistemi del mondo tolemaico, e copernico*. Florence; *Discorsi e dimostrazioni matematiche intorno a due nuove scienze attenenti alla meccanica* (*Dialogues concerning two new sciences*) (1638).
17. Gauss, C. F. (1801). *Disquisitiones Arithmeticae*.
18. Graunt, J. (1662). *Natural and political observations mentioned in a following index, and made upon the bills of mortality*. London.

© Springer Nature Switzerland AG 2018

P. N. Kaloyerou, *Basic Concepts of Data and Error Analysis*,
https://doi.org/10.1007/978-3-319-95876-7

19. Hacking, I. (2006). *The emergence of probability: A philosophical study of early ideas about probability, induction and statistical inference* (2nd ed.). Cambridge: Cambridge University Press.

20. Halley, E. (1693). An estimate of the degrees of mortality of mankind, drawn from curious tables of the births and funerals at the city of Breslau; with an attempt to ascertain the price of annuities upon lives. *Philosophical Transactions of the Royal Society, 17*, 596–610; 654–6.

21. Hayenes, W. M. (Editor-in-Chief). (2017). *CRC Handbook of chemistry and physics* (97th ed.). New York: CRC Press.

22. Huygens, C. (1657). *Ratiociniis in aleae ludo.* In F. van Schooten (ed.) *Exercitionum Mathematicorum.* Amsterdam. (The Dutch version, which was written earlier, is *Van Rekeningh in Spelen van Geluck*, Amsterdam (1660); *Traité de la Lumière: avec, un discours de la pesanteur* (*Treatise on Light*) (1690)).

23. Kolmogorov, A. N. (1933). *Grundbegriffe der wahrscheinlichkeitsrechnung*; English translation: *Foundations of the theory of probability* (1950).

24. Laplace, P. S., & Marquis de (1796). *Exposition du systéme du monde* (*The system of the world*); *Traité de mécanique céleste* (*Celestial mechanics*) was released in five volumes over the period 1798 and 1827; *Théorie analytique des probabilités* (*Analytic theory of probability*) (1812).

25. Legendre, A. -M. (1825–37). *Traité des fonctions elliptiques* (*Treatise on elliptic functions*); *Nouvelles méthodes pour la détermination des orbites des comètes* (*New methods for the determination of comet orbits*) (1806).

26. Leibniz, G. W. (1666). *De Arte Combinatoria* (*On the art of combination*); *Nova Methodus pro Maximis et Minimis* (*New method for the greatest and the least*) (1684).

27. Maxwell, J. C. (1872). *Theory of heat*; *Treatise on electricity and magnetism* (1873).

28. de Moivre, A. (1718). *The doctrine of chances, or a method of calculating the probability of events in play*; *Miscellanea Analytica* (*Analytical Miscellany*) (1730).

29. Newton, I. (1686). *Philosophiae Naturalis Principia Mathematica* (*Mathematical principles of natural philosophy*).

30. Pascal, B. (1963). *Oeuvres completes.* L. Lafuma (ed.). Paris; *Traité du triangle arithmétique avec quelques autres petits traités sur la même matiére* (1665).

31. Poisson, S. -D. (1811 and 1831). *Traité de mécanique (Treatise on mechanics)*; *Théorie nouvelle de l'action capillaire* (*A new theory of capillary action*); *Théorie mathématique de la chaleur* (*Mathematical theory of heat*) (1835); *Recherches sur la probabilité des jugements en matiére criminelle et en matiére civile (Research on the probability of criminal and civil verdicts)* (1837).

32. Ross, S. (2010). *A first course in probability* (8th ed.). New Jersey: Pearson.

33. Squires, G. L. (2001). *Practical physics* (4th ed.). Cambridge: Cambridge University Press.

34. Venn, J. (1866). *The logic of chance*. London. (Third edition of 1888 reprinted, New York, 1962).
35. Vernier, P. (1631). *La construction, l'usage, et les propriétés du quadrant nouveau de mathématiques* (*The construction, uses, and properties of a new mathematical quadrant*) (1631).

Appendix B
Tables

B.1 Base SI Units

This appendix and all appendices that follow have been compiled from the *CRC Handbook of Chemistry and Physics*.[1]

Physical Quantity	Symbol for Quantity	SI Base unit	Symbol for unit
length	l	metre	m
mass	m	kilogram	kg
time	s	second	s
electric current	I	ampere	A
temperature	T	kelvin	K
amount of substance	n	mole	mole
luminous intensity	I_v	candela	cd

[1] Editor-in-chief: W. M. Haynes, *CRC Handbook of Chemistry and Physics*, 97th edition, (CRC Press, New York, 2017).

© Springer Nature Switzerland AG 2018
P. N. Kaloyerou, *Basic Concepts of Data and Error Analysis*,
https://doi.org/10.1007/978-3-319-95876-7

B.2   Definition of the Base SI Units

Unit	Definition of unit
metre	The distance traveled by light in a vacuum in 1/299,792,458 of a second.
kilogram	The mass equal to the mass of the international prototype kilogram of platinum-iridium kept at the International Bureau of Weights and Measures in Sèvres, France.
second	The duration of 9,192,631,770 periods of the radiation corresponding to the transition between the hyperfine levels of the ground state of the cesium-133 atom.
ampere	The current that, if maintained in two wires placed one metre apart in a vacuum, produces a force of 2×10^{-7} newtons per metre of length.
kelvin	The fraction 1/273.16 of the thermodynamic temperature of the triple point of water.
mole	The amount of substance containing as many elementary entities of the substance as there are atoms in 0.012 kg of carbon-12.
candela	The intensity of radiation emitted perpendicular to a surface of 1/600,000 square metre of a blackbody at a temperature of freezing platinum at a pressure of 101,325 newtons per square meter.

B.3   Supplementary SI Units

Physical Quantity	Symbol for Quantity	SI Base unit	Symbol for unit
plane angle	θ	radian	rad
solid angle	Ω	steradian	sr

Unit	Definition of unit
radian	The angle subtended by an arc of a circle of length equal to its radius.
steradian	The solid angle subtended by a square area on the surface of a sphere with sides equal to the radius of the sphere.

B.4   Derived SI Units

Note that not all derived units have special names. For such units, under the 'Name' column we have given their definition in words. Units which are named usually take the name of their originator, e.g., the newton or the joule. Note that despite being named after a person, such units are **always** written in lowercase, i.e., the first letter is **not** capitalised. However, symbols for many such units are capitalised, e.g., the symbol for the newton is N, and the symbol for the joule is J.

Derived SI Units with Special names				
Physical Quantity	**Symbol for Quantity**	**Name of Unit**	**Symbol**	**Definition in Terms of Base Unit**
mass	m	atomic mass unit	u	$1.66053873 \times 10^{-27}$ kg
electric charge	C	coulomb	C	$A \cdot s$
electric potential difference, or voltage, or electromotive force	V	Volt	V	$W \cdot A^{-1} J = J \cdot C^{-1}$ $= J \cdot (A \cdot s)^{-1} =$ $kg \cdot m^2 s^{-3} \cdot A^{-1}$
electric resistance	R	ohm	Ω	$V \cdot A^{-1} = kg \cdot m^2 s^{-3} \cdot A^{-2}$
electric conductance	σ	siemens	S	$\Omega^{-1} = A \cdot V^{-1} =$ $s^3 \cdot A^2 kg^{-1} \cdot m^{-2}$
electric capacitance	C	farad	F	$CV^{-1} = A^2 \cdot s^4 \, kg^{-1} m^{-2} =$ $A \cdot s \, V^{-1} = C^2 \cdot N^{-1} \cdot m^{-1}$
electric inductance	L	henry	H	$kg \cdot m^2 \cdot s^{-2} A^{-2} =$ $V \cdot s \cdot A^{-1} = \Omega \cdot s$
energy	E	joule	J	$kg \cdot m^2 \cdot s^{-2} = N \cdot m$
force	F	newton	N	$kg \cdot m \cdot s^{-2} = J \cdot m^{-1}$
frequency	f	hertz	Hz	$cycles \cdot s^{-1} = s^{-1}$
heat	Q	joule	J	$kg \cdot m^2 \cdot s^{-2} = N \cdot m$
luminous flux	ϕ	lumen	lm	$cd \cdot sr$
Illumination or Illuminance	E	lux	lx	$cd \cdot sr \cdot m^{-2} = lm \cdot m^{-2}$
magnetic induction, magnetic flux density	B	tesla	T	$Wb \cdot m^{-2} = V \cdot s \cdot m^{-2} =$ $kg \cdot s^{-2} A^{-1}$
magnetic flux	Φ_B	weber	Wb	$V \cdot s = kg \cdot m^2 \cdot s^{-2} \cdot A^{-1}$
power	P	watt	W	$kg \cdot m^2 \cdot s^{-3} = J \cdot s^{-1}$
pressure	P	pascal	Pa	$kg \cdot m^{-1} \cdot s^{-2} = N \cdot m^{-2}$
temperature	T	degree kelvin	K	$0°C = 273.15$ K

Derived SI Units without Special names			
Physical Quantity	**Symbol for Quantity**	**Definition in Words in Terms of Base Unit**	**Definition in Terms of Base Units**
acceleration - linear	a	metres per second per second	$m \cdot s^{-2}$
acceleration - angular	α	radians per second per second	$rad \cdot s^{-2}$
area	A	metres-squared or square metres	m^2
density	ρ	kilograms per metre cubed	$kg \cdot m^{-3}$
current density	J	amps per meter squared	$A \cdot m^{-2}$
electric resistivity	ρ	ohm metre	$\Omega \cdot m = A^{-1}V \cdot m$
electric conductivity	σ	inverse ohm metre	$(\Omega \cdot m)^{-1} = A \cdot (V \cdot m)^{-1}$
electric dipole moment	p	coulomb metre	$C \cdot m$
electric field strength	E	newtons per coulomb or volts per metre	$N \cdot C^{-1} = V \cdot m^{-1}$
electric flux	Φ_E	newton metre squared per coulomb or volt metre	$N \cdot m^2 \cdot C^{-1} = V \cdot m$
entropy	S	joules per degree kelvin	$J \cdot K^{-1} = J \cdot (^\circ C)^{-1}$
magnetic dipole moment	μ	ampere metre squared	$A \cdot m^2$
magnetic field strength, or magnetic field intensity	H	amperes per metre	$A \cdot m^{-1}$
specific heat capacity	c	joules per kilogram per degree kelvin	$J \cdot (kg \cdot K)^{-1}$
molar heat capacity	C_V or C_P	joules per mole per degree kelvin	$J \cdot (mol \cdot K)^{-1}$
moment of force, or torque	τ	newton metre	$N \cdot m$
moment of inertia	I	kilogram meter squared	$kg \cdot m^2$
velocity - linear	v	metres per second	$m \cdot s^{-1}$
velocity - angular	ω	radians per second	$rad \cdot s^{-1}$
volume	V	metre cubed or cubic metre	m^3

B.5   Prefixes for Powers of 10

Prefix	Symbol	Value	Prefix	Symbol	Value
yocto	y	10^{-24}	deca	da	10^1
zepto	z	10^{-21}	hecto	h	10^2
atto	a	10^{-18}	kilo	k	10^3
femto	f	10^{-15}	mega	M	10^6
pico	p	10^{-12}	giga	G	10^9
nano	n	10^{-9}	tera	T	10^{12}
micro	μ	10^{-6}	peta	P	10^{15}
milli	m	10^{-3}	exa	E	10^{18}
centi	c	10^{-2}	zetta	Z	10^{21}
deci	d	10^{-1}	yotta	Y	10^{24}

B.6   Conversion Tables

Length	Pressure
1 in = 2.54 cm (exact)	1 Pa = 1 N·m^{-2}
1 m = 39.37 in = 3.281 ft	1 bar = 10^5 1 Pa
1 ft (foot) = 12 in = 0.3048 m	1 atm = 760 mm Hg = 76.0 cm Hg
1 yd (yard) = 3 ft = 0.9144 m	1 atm = 1.013×10^5 N·m^{-2}
1 km = 0.6214 mi	1 torr = 1 mm Hg = 133.3 Pa
1 mi (mile) = 1 mi = 5280 ft = 1.609 km	
1 Å(ångstrom) = 10^{-10} m	**Time**
1 lightyear = 9.461×10^{15} m	1 yr = 365.25 days = 3.156×10^7 s
	1 day = 24 h = 1.44×10^3 min = 8.64×10^4 s
Volume	
1 l (litre) = 1000 ml = 1000 cm^3	**Energy**
1 ml = 1 cm^3	1 cal = 4.186 J
1 gal (UK) = 4.546 l	1 J = 0.2389 cal
1 l = 0.2100 gal (UK)	1 Btu = 252 cal = 1054 J
1 gal (US) = 3.785 l	1 eV = 1.602×10^{-19} J
1 l = 0.2642 gal (US)	1 kWh = 3.600×10^6 J
Mass	**Power**
1 t (metric ton or tonne) = 1000 kg =	1 hp = 746 W
1 t (UK ton) = 1016 kg	1 W = 1 J·s^{-1}
1 slug = 14.59 kg	1 Btu·h^{-1} = 0.293 W
1 u = 1.661×10^{-27} kg = 931.5 MeV·c^{-2}	
	Temperature
Force	1 °C = 1 K
1N = 0.2248 lb	$T_F = \left(\frac{9}{5}T_C + 32\right)^{\circ}$F
1 lb = 4.448 N	$T_C = \frac{5}{9}(T_F - 32)^{\circ}$C

B.7 Physical Constants

The physical constants are given to four significant figures.

Quantity	Symbol/formula	Value
Atomic mass unit	u	1.661×10^{-27} kg $= 931.5$ MeV$\cdot$c^{-2}
Avogadro's number	N_A	6.022×10^{23} particles$\cdot$mol^{-1}
Bohr magneton	$\mu_B = e\hbar/(2m_e)$	9.274×10^{-24} J$\cdot$T^{-1}
Bohr radius	$a_0 = \hbar^2/(m_e e^2 k_e)$	5.292×10^{-11} m
Boltzmann's constant	$k_B = R/N_A$	1.381×10^{-23} J$\cdot$K^{-1}
Compton wavelength	$\lambda_C = h/(m_e c)$	2.426×10^{-12} m
Coulomb constant	$k_e = 1/(4\pi\epsilon_0)$	8.988×10^9 N$\cdot$m^2$\cdot$C^{-2}
Electron mass	m_e	9.109×10^{-31} kg $= 5.486 \times 10^{-4}$ u $= 0.5110$ MeV$\cdot$c^{-2}
Electron volt	eV	1.602×10^{-19} J
Elementary charge	e	1.602×10^{-19} C
Gas constant	R	8.314 J$\cdot$(K$\cdot$mol)$^{-1}$
Gravitational constant	G	6.673×10^{-11} N$\cdot$m^2$\cdot$kg^{-2}
Josephson frequency-voltage ratio	$2e/h$	4.836×10^{14} Hz$\cdot$V^{-1}
Magnetic flux quantum	$\Phi_0 = h/(2e)$	2.068×10^{-15} T$\cdot$m^2
Neutron mass	m_n	1.675×10^{-27} kg $= 1.009$ u $= 939.6$ MeV$\cdot$c^{-2}
Nuclear magneton	$\mu_n = e\hbar/(2m_p)$	5.051×10^{-27} J$\cdot$T^{-1}
Permeability of free space	μ_0	$4\pi \times 10^{-7}$ T$\cdot$m$\cdot$A^{-1}
Permitivity of free space	$\epsilon_0 = 1/(\mu_0 c^2)^{-1}$	8.854×10^{-12} C^2$\cdot$(N$\cdot$m^2)$^{-1}$
Planck's constant	h	6.626×10^{-34} J$\cdot$s^{-1}
	$\hbar/(2\pi)$	1.055×10^{-34} J$\cdot$s^{-1}
Proton mass	m_p	1.673×10^{-27} kg $= 1.007$ u $= 938.3$ MeV$\cdot$c^{-2}
Rydberg constant	R_H	1.097×10^7 m^{-1}
Speed of light in vacuum	c	2.998×10^8 m$\cdot$s^{-1}

Other Constants

Quantity	Symbol/formula	Value
Mechanical equivalent of heat		4.186 J$\cdot$cal^{-1}
Standard atmospheric pressure	1 atm	1.013×10^5 Pa
Absolute zero	0 K	-273.15 °C
Acceleration due to gravity	g	9.807 m$\cdot$s^{-2}

Planet	Mass kg	Mean Radius km	Mean Orbital Radius km	Orbital Period
Sun	1.99×10^{30}	6.96×10^5	−	−
Moon	7.35×10^{22}	1.74×10^3	3.84×10^5	27.3 d
Mercury	3.30×10^{23}	2.44×10^3	5.79×10^7	88.0 d
Venus	4.87×10^{24}	6.05×10^3	1.08×10^8	224.7 d
Earth	5.97×10^{24}	6.38×10^3	1.50×10^8	365.3 d
Mars	6.42×10^{23}	3.40×10^3	2.28×10^8	687.0 d
Jupitar	1.90×10^{27}	6.91×10^4	7.78×10^8	11.86 y
Saturn	5.68×10^{26}	6.03×10^4	1.43×10^9	29.45 y
Uranus	8.68×10^{25}	2.56×10^4	2.87×10^9	84.02 y
Neptune	1.02×10^{26}	2.48×10^4	4.50×10^9	164.8 y
Pluto	1.31×10^{22}	1.15×10^3	5.91×10^9	247.9 y

B.8   Solar System Data

B.9   The Greek Alphabet

alpha	A	α	beta	B	β	gamma	Γ	γ
delta	Δ	δ	epsilon	E	ϵ	zeta	Z	ζ
eta	H	η	theta	Θ	θ	iota	I	ι
kappa	K	κ	lambda	Λ	λ	mu	M	μ
nu	N	ν	xi	Ξ	ξ	omicron	0	o
pi	Π	π	rho	P	ρ	sigma	Σ	σ
tau	T	τ	upsilon	Υ	υ	phi	Φ	ϕ
chi	X	χ	psi	Ψ	ψ	omega	Ω	ω

Appendix C
Some Mathematical Formulae and Relations

(1) Laws of algebraic operations:

  (i) Commutative law:
$$a + b = b + a, \quad ab = ba,$$

  (ii) Associative law:
$$a + (b + c) = (a + b) + c, \quad a(bc) = (ab)c,$$

  (iii) Distributive law:
$$c(a + b) = ca + cb.$$

(2) Operation with fractions:
$$\frac{a}{b} \pm \frac{c}{d} = \frac{ad \pm bc}{bd}, \quad \frac{a}{b} \times \frac{c}{d} = \frac{ac}{bd}, \quad \frac{a}{b} \div \frac{c}{d} = \frac{a}{b} \times \frac{d}{c} = \frac{ad}{bc}.$$

(3) Laws of exponents:
$$a^m \times a^n = a^{m+n}, \quad (ab)^n = a^n b^n, \quad (a^m)^n = a^{mn}.$$

(4) Zero and negative exponents:
$$a^0 = 1 \text{ if } a \neq 0, \quad a^{-n} = \frac{1}{a^n}, \quad a^m \div a^n = a^{m-n}.$$

(5) Laws of base 10 logarithms:

© Springer Nature Switzerland AG 2018
P. N. Kaloyerou, *Basic Concepts of Data and Error Analysis*,
https://doi.org/10.1007/978-3-319-95876-7

$$\log a = x \Rightarrow a = 10^x,$$

$$\log a + \log b = \log ab, \qquad \log a - \log b = \log \frac{a}{b},$$

$$\log a^n = n \log a, \qquad \log a^{-n} = \log \frac{1}{a^n} = -n \log a,$$

$$\log a^{\frac{1}{n}} = \frac{1}{n} \log a.$$

(6) Laws of natural logarithms, i.e., base e logarithms:

$$\ln a = x \Rightarrow a = e^x,$$

$$\ln a + \ln b = \ln ab, \qquad \ln a - \ln b = \ln \frac{a}{b},$$

$$\ln a^n = n \ln a, \qquad \ln a^{-n} = \ln \frac{1}{a^n} = -n \ln a,$$

$$\ln a^{\frac{1}{n}} = \frac{1}{n} \ln a.$$

(7) Binomial theorem (n positive):

$$(a+b)^n = a^n + \binom{n}{1} a^{n-1}b + \binom{n}{2} a^{n-2}b^2 + \cdots + \binom{n}{n-1} ab^{n-1} + \binom{n}{n} b^n$$

$$= \sum_{i=0}^{n} \binom{n}{i} a^{n-i} b^i,$$

where the coefficients are given by

$$\binom{n}{r} = {}_nC_r = \frac{n!}{(n-r)!r!} = \frac{{}_nP_r}{r!}.$$

(8) Roots of a quadratic equation $ax^2 + bx + c = 0$ are

$$\frac{-b \pm \sqrt{b^2 - 4ac}}{2a}.$$

(9) Permutations are the number of ways of ordering objects when the order of the objects matters. For example, three objects labeled a, b and c can be ordered in 6 ways: abc, cab, bca, bac, acb and cba. Two general permutations are as follows:

  (i) The number of ways of arranging n objects is $n!$,
  (ii) The number of ways of arranging r objects chosen from n objects is

$$_nP_r = \frac{n!}{(n-r)!}.$$

(10) Combinations are the number of ways of arranging n objects when the order does not matter. In this case the 6 ways of permuting three objects a, b and c are viewed as a single combination. The number of ways of arranging r objects chosen from n objects when the order objects does not matter is

$$\binom{n}{r} = {}_nC_r = \frac{n!}{(n-r)!r!} = \frac{{}_nP_r}{r!}.$$

(11) Geometry

   (i) Area A and perimeter P of a rectangle of length a and width b: $A = ab$, $P = 2a + 2b$,

   (ii) Area A and perimeter P of a parallelogram with acute angle θ, long sides a, short sides b and height $h = b \sin \theta$: $A = ah = ab \sin \theta$, $P = 2a + 2b$,

   (iii) Area A and perimeter P of a triangle of base a, sides b and c, and height $h = b \sin \theta$, where θ is the angle between b and a: $A = \frac{1}{2}ah = \frac{1}{2}ab \sin \theta$, $P = a + b + c$,

   (iv) Area A and perimeter P of a trapezoid of height h, parallel sides a and b, and acute angles θ and ϕ: $A = \frac{1}{2}h(a + b)$, $P = a + b + h\left(\frac{1}{\sin \theta} + \frac{1}{\sin \phi}\right)$,

   (v) Area A and circumference C of a circle of radius r: $A = \pi r^2$, $C = 2\pi r$,

   (vi) Area A of an ellipse with short radius (semi-minor axis) b and long radius (semi-major axis) a: $A = \pi ab$,

   (vii) Volume V and surface area A of a sphere of radius r: $V = \frac{4}{3}\pi r^3$, $A = 4\pi r^2$,

   (viii) Volume V and surface area A of a cylinder of radius r and height h: $V = \pi r^2 h$, $A = 2\pi rh + 2\pi r^2$,

   (ix) Volume V and surface area A of a rectangular box of length a, height b and width c: $V = abc$, $A = 2(ab + ac + bc)$,

   (x) Volume V of a parallelepiped with the rectangular face of area $A = ac$, and the parallelogram face with parallel sides a and b, acute angle θ and height (distance between parallel sides a) $h = b \sin \theta$: $V = Ah = acb \sin \theta$,

   (xi) Volume V and surface area A of a right circular cone of base radius r and height h: $V = \frac{1}{3}\pi r^2 h$, $A = \pi r^2 + \pi r \sqrt{r^2 + h^2}$,

   (xi) Volume V of a pyramid of base area A and height h: $V = \frac{1}{3}Ah$.

(12) Definition of a radian. A *radian* is defined as the angle subtended at the center of a circle by an arc equal to its radius. It is abbreviated as "rad":

$$1 \text{ rad} = \frac{360}{2\pi} = 57.30° \text{ (to four significant figures).}$$

(13) Trigonometric functions: sine (sin), cosine (cos) and tangent (tan):

   (i) Consider a right angled triangle with hypotenuse H, adjacent A and opposite O with respect to angle θ. Then, the trigonometric functions are defined as follows:

$$\cos\theta = \frac{A}{H}, \quad \text{cosecant} = \csc\theta = \frac{1}{\cos\theta}, \quad \text{inverse of } \cos\theta = \arccos\theta = \cos^{-1}\theta$$

$$\sin\theta = \frac{O}{H}, \quad \text{secant} = \sec\theta = \frac{1}{\sin\theta}, \quad \text{inverse of } \sin\theta = \arcsin\theta = \sin^{-1}\theta$$

$$\tan\theta = \frac{O}{A}, \quad \text{cotangent} = \cot\theta = \frac{1}{\tan\theta}, \quad \text{inverse of } \tan\theta = \arctan\theta = \tan^{-1}\theta$$

$$\tan\theta = \frac{\sin\theta}{\cos\theta}, \quad \sin^2\theta + \cos^2\theta = 1, \quad \sec^2\theta - \tan^2\theta = 1, \quad \csc^2\theta - \cot^2\theta = 1,$$

$$\cos(-\theta) = \cos\theta, \quad \sin(-\theta) = -\sin\theta, \quad \tan(-\theta) = -\tan\theta.$$

(ii) Addition formulae:

$$\sin(\theta \pm \phi) = \sin\theta\cos\phi \pm \sin\phi\cos\theta,$$
$$\cos(\theta \pm \phi) = \cos\theta\cos\phi \mp \sin\theta\sin\phi,$$
$$\tan(\theta \pm \phi) = \frac{\tan\theta \pm \tan\phi}{1 \mp \tan\theta\tan\phi}.$$

(iii) Double angle formulae:

$$\sin 2\theta = 2\sin\theta\cos\theta,$$
$$\cos 2\theta = \cos^2\theta - \sin^2\theta = 1 - 2\sin^2\theta = 2\cos^2\theta - 1,$$
$$\tan 2\theta = \frac{2\tan\theta}{1 - \tan^2\theta}.$$

(14) Hyperbolic functions:

(i)

$$\sinh\theta = \frac{e^\theta - e^{-\theta}}{2}, \quad \cosh\theta = \frac{e^\theta + e^{-\theta}}{2}, \quad \tanh\theta = \frac{e^\theta - e^{-\theta}}{e^\theta + e^{-\theta}} = \frac{\sinh\theta}{\cosh\theta}$$

$$\sinh(-\theta) = -\sinh\theta, \quad \cosh(-\theta) = \cosh\theta, \quad \tanh(-\theta) = -\tanh\theta.$$

(ii) Addition formulae:

$$\sinh(\theta \pm \phi) = \sinh\theta\cosh\phi \pm \sinh\phi\cosh\theta,$$
$$\cosh(\theta \pm \phi) = \cosh\theta\cosh\phi \pm \sinh\theta\sinh\phi,$$
$$\tanh(\theta \pm \phi) = \frac{\tanh\theta \pm \tanh\phi}{1 \pm \tanh\theta\tanh\phi}.$$

(iii) Double angle formulae:

$$\sinh 2\theta = 2\sinh\theta\cosh\theta,$$

$$\cosh 2\theta = \cosh^2\theta + \sinh^2\theta = 1 + 2\sinh^2\theta = 2\cosh^2\theta - 1,$$

$$\tanh 2\theta = \frac{2\tanh\theta}{1+\tanh^2\theta}.$$

(15) Relationship between hyperbolic and trigonometric functions:

$$\sin ix = i\sinh x, \qquad \cos ix = \cosh x, \qquad \tan ix = i\tanh x,$$

$$\sinh ix = i\sin x, \qquad \cosh ix = \cos x, \qquad \tanh ix = i\tan x.$$

(16) Series

$$e^x = 1 + x + \frac{x^2}{2!} + \frac{x^3}{3!} + \cdots \qquad\qquad -\infty < x < \infty,$$

$$e^{kx} = 1 + kx + \frac{k^2 x^2}{2!} + \frac{k^3 x^3}{3!} + \cdots \qquad -\infty < x < \infty,$$

$$\sin x = x - \frac{x^3}{3!} + \frac{x^5}{5!} - \frac{x^7}{7!} + \cdots \qquad -\infty < x < \infty,$$

$$\cos x = 1 - \frac{x^2}{2!} + \frac{x^4}{4!} - \frac{x^6}{6!} \cdots \qquad -\infty < x < \infty,$$

$$\tan x = x + \frac{x^3}{3} + \frac{2x^5}{15} + \frac{17x^7}{315} \cdots \qquad |x| < \frac{\pi}{2},$$

$$\sinh x = x + \frac{x^3}{3!} + \frac{x^5}{5!} + \frac{x^7}{7!} + \cdots \qquad -\infty < x < \infty,$$

$$\cosh x = 1 + \frac{x^2}{2!} + \frac{x^4}{4!} + \frac{x^6}{6!} \cdots \qquad -\infty < x < \infty,$$

$$\tanh x = x - \frac{x^3}{3} + \frac{2x^5}{15} - \frac{17x^7}{315} \cdots\cdots \qquad |x| < \frac{\pi}{2}.$$

(17) Definition of the derivative of a function $u(x)$:

$$\frac{du}{dx} = \lim_{\Delta x \to 0} \frac{u(x+\Delta x) - u(x)}{\Delta x}.$$

(18) Product rule. If $f = f(x)$ and $g = g(x)$ then the product rule gives

$$\frac{dfg}{dx} = f\frac{dg}{dx} + \frac{df}{dx}g.$$

(19) Chain rule:

(a) The chain rule for the derivative of $f = f[u(x)]$ is

$$\frac{df}{dx} = \frac{df}{du}\frac{du}{dx}.$$

(b) The chain rule for the derivative of $f = f[u(x), v(x)]$ is

$$\frac{df}{dx} = \frac{df}{du}\frac{du}{dx} + \frac{df}{dv}\frac{dv}{dx}.$$

(20) Derivatives. In this section c is a constant and $u = u(x)$ is a function of x.

$$\frac{dc}{dx} = 0, \qquad \frac{d(cu)}{dx} = c\frac{du}{dx},$$

$$\frac{dx^n}{dx} = nx^{n-1}, \qquad \frac{du^n}{du} = nu^{n-1}\frac{du}{dx},$$

$$\frac{de^x}{dx} = e^x, \qquad \frac{de^u}{dx} = e^u\frac{du}{dx},$$

$$\frac{d\log_a x}{dx} = \frac{\log_a e}{x}, \qquad \frac{d\log_a u}{dx} = \frac{\log_a e}{u}\frac{du}{dx}, \qquad (a \neq 0, 1),$$

$$\frac{d\ln x}{dx} = \frac{1}{x}, \qquad \frac{d\ln u}{dx} = \frac{1}{u}\frac{du}{dx},$$

$$\frac{d\sin x}{dx} = \cos x, \qquad \frac{d\sin u}{dx} = \cos u\frac{du}{dx},$$

$$\frac{d\cos x}{dx} = -\sin x, \qquad \frac{d\cos u}{dx} = -\sin u\frac{du}{dx},$$

$$\frac{d\tan x}{dx} = \sec^2 x, \qquad \frac{d\tan u}{dx} = \sec^2 u\frac{du}{dx},$$

$$\frac{d\arcsin x}{dx} = \frac{1}{\sqrt{1-x^2}} \quad \left[-\frac{\pi}{2} < \arcsin x < \frac{\pi}{2}\right],$$

$$\frac{d\arcsin u}{dx} = \frac{1}{\sqrt{1-u^2}}\frac{du}{dx} \quad \left[-\frac{\pi}{2} < \arcsin u < \frac{\pi}{2}\right],$$

$$\frac{d\arccos x}{dx} = \frac{-1}{\sqrt{1-x^2}} \quad [0 < \arccos x < \pi],$$

$$\frac{d\arccos u}{dx} = \frac{-1}{\sqrt{1-u^2}}\frac{du}{dx} \quad [0 < \arccos u < \pi],$$

$$\frac{d\arctan x}{dx} = \frac{1}{1+x^2} \quad \left[-\frac{\pi}{2} < \arctan x < \frac{\pi}{2}\right],$$

$$\frac{d\arctan u}{dx} = \frac{1}{1+u^2}\frac{du}{dx} \quad \left[-\frac{\pi}{2} < \arctan u < \frac{\pi}{2}\right],$$

$$\frac{d\sinh x}{dx} = \cosh x, \qquad \frac{d\sinh u}{dx} = \cosh u\frac{du}{dx},$$

$$\frac{d\cosh x}{dx} = \sinh x, \qquad \frac{d\cosh u}{dx} = \sinh u\frac{du}{dx},$$

$$\frac{d\tanh x}{dx} = \operatorname{sech}^2 x, \qquad \frac{d\tanh u}{dx} = \operatorname{sech}^2 u\frac{du}{dx}.$$

(21) Indefinite integrals. In this section c and n are constants, while $u = u(x)$ and $v = v(x)$ are functions of x.

$$\int 0 \, dx = 0, \qquad \int dx = x + c, \qquad \int cu \, dx = c \int u \, dx,$$

$$\int x^n \, dx = \frac{x^{n+1}}{n+1} \quad (n \neq -1), \qquad \int e^x \, dx = e^x + c,$$

$$\int \frac{1}{x} \, dx = \ln x + c \ \ (x > 0), \qquad \int \frac{1}{x} \, dx = \ln(-x) + c \ \ (x < 0),$$

$$\int a^x \, dx = \int e^{x \ln a} \, dx = \frac{e^{x \ln a}}{\ln a} + c = \frac{a^x}{\ln a} + c, \qquad a > 0, a \neq 1,$$

$$\int \sin x \, dx = -\cos x + c, \qquad \int \cos x \, dx = \sin x + c,$$

$$\int \tan x \, dx = \ln \sec x + c = -\ln \cos x + c,$$

$$\int \sinh x \, dx = \cosh x + c, \qquad \int \cosh x \, dx = \sinh x + c,$$

$$\int \tanh x \, dx = \ln \cosh x + c,$$

(22) Integration by parts. Let $u = u(x)$ and $v = v(x)$ be functions of x.

$$\int \frac{du}{dx} v \, dx = uv - \int u \frac{dv}{dx} \, dx, \qquad \int_a^b \frac{du}{dx} v \, dx = [uv]_a^b - \int u \frac{dv}{dx} \, dx$$

(23) Definite integrals: In this section m and n are integers.

$$\int_0^\infty \frac{1}{x^2 + a^2} \, dx = \frac{\pi}{2a}, \qquad \int_0^\infty \frac{x^{p-1}}{1+x} \, dx = \frac{\pi}{\sin p\pi} \quad (0 < p < 1),$$

$$\int_0^\infty \frac{1}{\sqrt{a^2 - x^2}} \, dx = \frac{\pi}{2}, \qquad \int_0^\infty \sqrt{a^2 - x^2} \, dx = \frac{\pi a^2}{4},$$

$$\int_0^\pi \sin mx \sin nx \, dx = \begin{cases} 0 & m \neq n \\ \frac{\pi}{2} & m = n \end{cases},$$

$$\int_0^\pi \cos mx \cos nx \, dx = \begin{cases} 0 & m \neq n \\ \frac{\pi}{2} & m = n \end{cases},$$

$$\int_0^\pi \sin mx \cos nx \, dx = \begin{cases} 0 & m + n \text{ even} \\ \frac{2m}{m^2 - n^2} & m + n \text{ odd} \end{cases},$$

$$\int_0^\infty \frac{\sin px}{x} \, dx = \begin{cases} \frac{\pi}{2} & p > 0 \\ 0 & p > 0 \\ -\frac{\pi}{2} & p < 0 \end{cases}$$

$$\int_0^\infty x^n e^{-ax} = \frac{\Gamma(n+1)}{a^{n+1}}, \quad n \text{ integer}, \quad \text{or} \quad = \frac{n!}{a^{n+1}}, \quad n = 0, 1, 2, \ldots$$

$$\int_0^\infty e^{-ax} \cos bx\, dx = \frac{a}{a^2+b^2}, \qquad \int_0^\infty e^{-ax} \sin bx\, dx = \frac{b}{a^2+b^2},$$

$$\int_0^\infty e^{-ax^2}\, dx = \frac{1}{2}\sqrt{\frac{\pi}{a}}, \qquad \int_0^\infty x^{2n} e^{-ax^2}\, dx = \frac{1 \cdot 3 \cdot 5 \cdots (2n-1)}{2^{n+1} a^n}\sqrt{\frac{\pi}{a}},$$

$$\int_0^\infty e^{-x^2-\frac{a^2}{x^2}}\, dx = \frac{1}{2} e^{-2a}\pi^{\frac{1}{2}}, \qquad \int_0^\infty e^{-ax^2} \cos bx\, dx = \frac{1}{2}\sqrt{\frac{\pi}{a}} e^{-b^2/4a},$$

$$\int_0^\infty e^{-x} \ln x\, dx = -\gamma, \qquad \int_0^\infty e^{-x^2} \ln x = -\frac{\sqrt{\pi}}{4}(\gamma + 2\ln 2),$$

where $\gamma = 0.5772\ldots$ is Euler's constant

(24) The exponential function as the limit of an infinite product

$$\lim_{n\to\infty}\left(1+\frac{x}{n}\right) = e^x, \qquad \lim_{n\to\infty}\left(1-\frac{x}{n}\right) = e^{-x}$$

(25) The gamma function, $\Gamma(x)$, for x a real positive number is defined by

$$\Gamma(x) = \int_0^\infty e^{-t} t^{x-1}\, dt, \qquad x > 0$$

The gamma function for arbitrary x is defined by

$$\Gamma(x) = \lim_{n\to\infty}\frac{n! n^{x-1}}{x(x+1)(x+2)\cdots(x+n-1)}, \qquad x \text{ arbitrary}$$

For negative real x the gamma function has the following values

$$\Gamma(x) = \begin{cases} \pm \text{ finite real value for negative } x \neq -1, -2, \ldots \\ \pm \text{ for } x = -1, -2, \ldots \end{cases}$$

Some properties of the gamma function

(i) $\Gamma(x+1) = x\Gamma(x), \qquad$ real $x > 0$

(ii) $\Gamma(n+1) = n!, \qquad$ positive integerl n

(iii) $\Gamma(x)\Gamma(1-x) = \dfrac{\pi}{\sin \pi x}$

(26) The factorial function, $\Pi(n) = n!$, for positive integer n is defined by

$$\Pi(n) = n! = n(n-1)(n-2)\ldots 1$$

Using the formula $\Pi(n) = \Gamma(n + 1)$, the definition of the factorial function can be extended to 0 and negative integer n as follows:

$$\Pi(0) = 0! = 1$$
$$\Pi(n) = \pm\infty, \qquad n = \text{negative integer}$$

(27) Stirling's approximation to $n!$

$$n! = \sqrt{2\pi n}\, n^n e^{-n}$$

Note, this formula was actually first introduced by de Moivre.

(28) Some important numbers

$$\pi = 3.14159265358979323846264338327950288419716939937511$$

Euler's constant γ:

$$\gamma = 0.57721566490153286061$$

Base of natural numbers e:

$$e = 2.71828182845904523536028747135266249775724709369996.$$

Appendix D
Some Biographies

- **Aristotle**, 384–322, Ancient Greek philosopher and scientist. He was born in Stagira, Chalcidice, Greece. His father was the physician of the king of Macedonia. He later moved to Athens, where he joined the Academy of Plato, remaining there for the next twenty years. Aristotle studied many areas of science and the arts including biology, botany, chemistry, physics, logic, ethics, political theory, poetics and philosophy. His philosophical work is relevant even to this day, and he is regarded as one of the great thinkers of his time. Aristotle wrote extensively and we mention here just some of his main works: *Peri ta zōa historiai* (*History of animals*), *Physikē* (*Physics*), *Peri psychēs* (*On the Soul*), *Ēthika Nikomacheia* (*Nicomachean Ethics*) and *Politika* (*Politics*).
- **Bernoulli, Daniel**, 1700–1782, Swiss mathematician. D. Bernoulli was born in Groningen, Netherlands to a family of mathematicians. Both his father, Johann, and his prominent uncle, Jacques Bernoulli, were mathematicians. He studied philosophy, logic and medicine at the Universities of Heidelberg, Strasbourg, and Basel. He won a position at the acclaimed Academy of Sciences in St. Petersburg, Russia, where he lectured in medicine, mechanics and physics until in 1732 he accepted a position in anatomy and botany at the University of Basel. Despite his most noted original contributions being in physics, he only took up a physics position in 1750. He worked in the areas of astronomy, gravity, tides, magnetism, ocean currents and hydrodynamics. He studied the behaviour of ships at sea and made significant contributions to probability in connection with his work on the kinetic theory of gases. He explained pressure, for example, as the impact of many randomly moving molecules. Perhaps, his most prominent work is *Hydrodynamica* published in 1738, which includes his famous Bernoulli equation which asserts that at any point in a liquid pressure + kinetic energy per unit volume + potential energy per unit volume are equal to the same constant.
- **Bernoulli, Jacques (Jakob)**, 1655–1705, Swiss mathematician. He was born in Basel, Switzerland, to a family of drug merchants. Bernoulli became professor of mathematics at the University of Basel in 1687. His book *Ars conjectandi* (*The Art of Conjecturing*), published posthumously in 1713, contains some of his finest work: the theory of permutations and combinations, the now-called Bernoulli

© Springer Nature Switzerland AG 2018
P. N. Kaloyerou, *Basic Concepts of Data and Error Analysis*,
https://doi.org/10.1007/978-3-319-95876-7

numbers with which he derived the exponential series, and early concepts of probability.

- **Boltzmann, Ludwig Eduard**, 1844–1906, Austrian physicist. Boltzmann was born in Vienna, Austria. He obtained his doctorate from the university of Vienna in 1866. He held professorial positions in mathematics and physics at Vienna, Graz, Munich, and Leipzig. His major contributions were in statistical mechanics. Among his contributions to statistical mechanics is the derivation of the principle of equipartition of energy and what is now called the Maxwell–Boltzmann distribution law.
- **Fermat, Pierre de**, 1601–1665, French mathematician. He was born in Beaumont-de-Lomagne, France, and educated at a Franciscan school, studying law in later years. Little is known of his early life. Fermat received the baccalaureate in law from the University of Orléans in 1631 and went on to serve as a councillor in the local parliament of Toulouse in 1634. Fermat made contributions to number theory, analytic geometry and probability. He is regarded as the inventor of differential calculus through his method of finding tangents and maxima and minima. Some 30 years later, Sir Isaac Newton introduced his calculus. Recognition of Fermat's work was scanty, perhaps because he used an older clumsy notation rendered obsolete by Descarte's 1637 work *Géométrie*.

  In his work reconstructing the long lost *Plane Loci* of the 3rd century BC Greek geometer Apollonius, Fermat found that the study of geometry is facilitated by the introduction of a coordinate system. Fermat's *Introduction to Loci* was published much later, after his death, in 1679. Descarte introduced a similar idea in his 1637 *Géométrie*. Since then, the study of geometry using a coordinate system has become known as Cartesian geometry, after René Descarte, whose name in Latin translates to Renatus Cartesius.

  Fermat differed with Descarte's attempt to explain the sine law of refraction (at a surface separating materials of different densities, the ratio of the sine of the angle of incidence and the angle of refraction is constant) by supposing that light travels faster in the denser medium. Fermat, on the other hand, influenced by the Aristotelian view that nature always chooses the shortest path, supposed that light travels faster in the less dense medium. Fermat was proved correct by the later wave theory of Huygens and by the 1849 experiment of Fizeau.
- **Galileo Galilei**, 1564–1642, Italian natural philosopher, astronomer and mathematician. He was born in Pisa, Tuscany, Italy, the oldest son of Vincenzo Galilei, a musician who made important contributions to the theory and practice of music. The Family moved to Florence, where Galileo attended the monastery school at Vallombrosa. He enrolled at the University of Pisa to study medicine, but instead, was captivated by mathematics and philosophy and switched to these subjects against his father's wishes. However, in 1585, he left the university without a degree. Despite not having a degree, his scientific and mathematical work gained recognition and in 1589 he was awarded the chair of mathematics at the University of Pisa. While at Pisa he performed his famous experiment in which he dropped

objects of different weights from the top of the Leaning Tower of Pisa, demonstrating that objects of different weights fall at the same rate, contrary to Aristotle's claim. His studies on the motion of bodies led him away from Aristotelian notions about motion, preferring instead the Archimedean approach.

In 1609, after hearing of the development of a new instrument, the telescope, Galileo soon discovered the secret of its construction and built a number of increasingly improved telescopes culminating in a telescope that could magnify 20 times. With this telescope he studied the heavens, confirming to himself the correctness of the heliocentric system of Copernicus (Nicolaus Copernicus, Polish astronomer, 1473–1543). Copernicus developed the heliocentric system sometime between 1508 and 1514, with the final version presented in his book *De revolutionibus orbium coelestium libri vi* (*Six Books Concerning the Revolutions of the Heavenly Orbs*) published in 1543, the year of his death. Expressing support for the Copernican system brought Galileo into conflict with the Catholic church, eventually resulting in the Inquisition (a Catholic judicial body mandated to prosecute heresy) sentencing him to life imprisonment. It should be noted that Galileo never spent time in a dungeon, instead, he served out his sentence in a variety of comfortable surroundings.

Galileo made important contributions to the sciences of motion (for example, he established that objects of different weights fall at the same rate), astronomy (providing evidence for the Copernican system to name one contribution to astronomy) and the strength of materials. Of particular importance is his development of the modern scientific method, where instead of investigating nature by pure thought, he maintained that facts about nature should be established by experiment and expressed in mathematical language. When about 70 years of age, Galileo completed his book on the science of motion and on the strength of materials entitled *Discorsi e dimostrazioni matematiche intorno a due nuove scienze attenenti alla meccanica* (*Dialogues concerning two new sciences*). The book was discretely taken out of Italy and published in Leiden, Netherlands in 1638.

- **Gauss, Carl Friedrich** - see footnote 1, Chap. 4.
- **Gibbs, Josiah Willard**, 1839–1903, American physicist and chemist. He was born in New Haven, Connecticut, U.S.A, the fourth child and only son of Josiah Willard Gibbs Sr., professor of sacred literature at Yale University. He entered Yale himself in 1854 as a student and went on to become a professor of mathematical physics at Yale in 1871 and remained there for the rest of his life. Gibbs was regarded as one of the greatest American scientists of his time. Gibbs made significant contributions to thermodynamics and statistical mechanics and is credited with converting physical chemistry from an empirical into a deductive science. One notable contribution to thermodynamics was the development in 1873 of a geometrical method for representing thermodynamic properties of substances by means of surfaces. James Clerk Maxwell, in England, impressed with Gibbs work, constructed a model of Gibbs's thermodynamic surface and sent it to him.
- **Huygens, Christiaan**, 1629–1695, Dutch mathematician, astronomer and physicist. Huygens was born in the Hague, the Netherlands, to a wealthy family. His

father, Constantijn Huygens, was a diplomat and dabbled in poetry. From an early age, Huygens showed a talent for drawing and mathematics. He became a student at the University of Leiden, where he studied mathematics and law.

Huygens improved the construction of telescopes using his new method of grinding and polishing lenses. This allowed him, in 1654, to identify that the funny shape of Saturn observed by Galileo was actually rings that circled the planet. His interest in astronomy required an accurate way to measure time and this led him to invent the pendulum clock. Huygens also made contributions to dynamics: he derived the formula for the time of oscillation of a simple pendulum, the oscillation of a body about a stationary axis, and the laws of centrifugal force for uniform circular motion (now described in terms of centripetal force) and in 1656 obtained solutions for the collision of elastic bodies (published later in 1669).

Huygens, however, remains most famous as the founder of the wave theory of light, presented in his book *Traité de la Lumière* (*Treatise on Light*). Though largely completed by 1678, it was only published in 1690. As made clear in his book, Huygens held the view that natural phenomena such as light and gravity are mechanical in nature and hence should be described by a mechanical model. This view led him to criticise Newton's theory of gravity. In describing light as waves, he was also in strong opposition to Newton corpuscular view of light (i.e., that light is made up of particles). Though requiring an underlying mechanical medium producing light waves, Huygens gave a beautiful description of reflection and refraction based on what is now called Huygens' principle of secondary wave fronts, which is a completely non-mechanical description.

- **Joule, James Prescott** - see footnote 3, Chap. 1.
- **Kelvin, Lord** - see footnote 1, Chap. 1.
- **Laplace, Pierre Simon Marquis de**, 1749–1827, French mathematician, astronomer and physicist. Laplace was born in Beaumount-en-Auge, Normandy, France, the son of a peasant farmer. At the military academy at Beaumont, he quickly showed his mathematical ability. In 1766 Laplace began studies at the University of Caen, but later left for Paris, apparently, before completing his degree. He took with him a letter of recommendation which he presented to Jean d'Alembert, who helped him find employment as a professor at the École Militaire, where he taught from 1769 to 1776. In later life, Laplace became president of the Board of Longitude, contributed to the development of the metric system and served for six weeks as minister of the interior under Napoleon. For his contributions, Laplace was eventually created a marquis (a nobleman ranking above a count but below a duke). He survived the French Revolution when many high-ranking individuals were executed.

Laplace made significant contributions to astronomy, mathematics and probability. Among his contributions was his highly regarded solution of the perplexing problem of the stability of the solar system, which he explained by his discovery that the average angular velocity of planets remained constant. Among his achievements in mathematics is the development of what are now called the Laplace equation and Laplace transforms that find application in many areas of physics and engineering. In probability theory, he developed numerous tools for calculating probabilities and showed that large amounts of astronomical data can be approximated by a Gaussian distribution.

In 1796 Laplace published *Exposition du systéme du monde* (*The System of the World*), describing his work on celestial mechanics for a general audience. In it, he postulated that the solar system originated from the cooling and contracting of a gaseous nebula. His vast and comprehensive work *Traité de mécanique céleste* (*Celestial mechanics*) was released in five volumes over the period 1798 and 1827. In this work, he applied the law of gravitation to the motion of planets and established a fairly complete model of the solar system. He also resolved a number of problems concerning tidal motion. Laplace's influential text *Théorie analytique des probabilités* (*Analytic theory of probability*), first published in 1812, included the contributions to probability mentioned above and a special case of the central limit theory. The introduction to this text was published for a general readership in 1814.

- **Legendre** - see footnote 2, Chap. 4.
- **Leibnitz, Gottfried Wilhelm Freiherr von**, 1646–1716, German philosopher, mathematician and political adviser. Leibnitz was born in Leipzig, Germany, to a religious (Lutheran) family. It was the time of the Thirty Years' War, which had left Germany in ruins. He attended Nicolai School, but was largely self taught in his father's large library. In 1661, he began a law degree at the University of Leipzig. Though his studies were in law, whilst at the University of Leipzig, he encountered the work of the great philosophers, mathematicians and physicists of his and earlier times and developed a life-long interest in these fields.

Following his studies, Liebniz first found employment as a lawyer. In subsequent years he held various positions, none of which were academic. In his final position, he was named historian for the House of Brunswick and continued his duties there until the end of his life.

However, throughout the period of his non-academic positions, Liebnitz continued his scientific and mathematical work. By late 1675 he introduced what might be regarded as his main contribution, the development of the foundations of both integral and differential calculus, published in a more complete form in his 1684 book *Nova Methodus pro Maximis et Minimis* (*New Method for the Greatest and the Least*). In 1676 he developed the new discipline of dynamics in which he replaced the conservation of movement by kinetic theory. In 1679 he perfected a binary system of arithmetic (arithmetic to base 2) and later the same year introduced early ideas on topology. He also worked on numerous engineering projects. One of these projects was the development of a water pump run by windmills which proved useful in mining in the Harz Mountains, where he frequently worked as an engineer between 1680 to 1685. Because of his rock observations during this time, he is considered among the creators of geology. He suggested that the earth was once molten.

- **Maxwell, James Clerk**, 1831–1879, Scottish physicist. Maxwell was born in Edinburgh to a well-off middle-class family. The original family name was Clerk, but his father, a lawyer, added the name Maxwell when he inherited Middlebie estate from Maxwell ancestors. At school age, he attended the Edinburgh Academy and when 16 years of age entered the University of Edinburgh. He later studied at the university of Cambridge where he excelled. In 1856 he was appointed to the professorship of natural philosophy at Marischal College, Aberdeen. When, in 1860, Marischal College merged with King's College to form the university of Aberdeen, Maxwell was made redundant. After his application to the university of Edinburgh was rejected, he managed to secure a position as professor of natural philosophy at King's College, London.

The five years following his appointment in King's College in 1860 were, perhaps, his most prolific. During this period he published two classic papers on the electromagnetic field, supervised the experimental determination of electrical units, and confirmed experimentally the speed of light predicted by his theory. In 1865 Maxwell resigned his professorship at King's College and retired to the family estate in Glenlair, where he devoted most his time to writing his famous treatise on electricity and magnetism *Treatise on Electricity and Magnetism* published in 1873. By unifying the experimentally established laws of electricity and magnetism (which includes Faraday's law of induction), he established his enormously powerful electromagnetic theory based on the four equations now known as Maxwell's equations. From these equations he derived a wave equation describing light, which established light as an electromagnetic wave. Electromagnetic waves have a broader spectrum than visible light and the connection of all such waves with electricity and magnetism suggested that these waves could be produced in the laboratory. This was confirmed experimentally eight years after Maxwell's death by Heinrich Hertz in 1887, when he succeed in producing radio waves (giving rise to the radio industry).

But this was not his only achievement. He made significant contributions to thermodynamics (the Maxwell relations) and statistical mechanics where, to mention one contribution, he derived a distribution law for molecular velocities, now called the Maxwell-Boltzmann distribution law. Maxwell, though a great theoretician, also possessed experimental skills and designed a number of experiments to investigate colour. This led him to suggest that a colour photograph could be produced using filters of the three primary colours (red, blue and green). He confirmed his proposal in an 1861 lecture to the Royal Institution of Great Britain, where he projected, through filters, a colour photograph of a tartan ribbon that he made by his method. His introduction of a hypothetical super being, now called Maxwell's Demon, played an important role in conceptual discussions of statistical ideas. His work on geometric optics led to the development of the fish-eye lens. His contributions to heat were published in his book *Theory of Heat* published in 1872.

Maxwell returned to an academic position in 1871, when he accepted his election to the new Cavendish professorship at Cambridge. During his tenure, he designed and supervised the construction of the Cavendish Laboratory.

- **Moivre, Abraham de**, 1667–1754, French mathematician. De Moivre was born in Vitry, France, to a protestant family and was later jailed for being a protestant following the revocation of the Edict of Nantes in 1685. When released shortly after, he moved to London, England, becoming friends with Sir Isaac Newton and the astronomer Edmond Halley[2]. Through his exceptional contributions to analytic trigonometry and to the theory of probability, De Moivre was elected to the Royal Society of London in 1967. Some time later he was also elected to to the Berlin and Paris academies. Despite his contributions to mathematics he was never appointed to a permanent academic position, instead earning a living as a tutor and a consultant to gamblers, working from a betting shop in Long Acres, London, known as Slaughter's Coffee House.

  In 1730, de Moivre introduced what is now called the Gaussian or normal distribution, which plays a central role in probability and statistics. The true significance of the distribution was realised much later when Gauss used it as a central part of his method for locating astronomical objects. As a result, it came to be known as the Gaussian distribution. So many sets of data satisfied the Gaussian distribution that it came be thought of as normal for a data set to satisfy the Gaussian curve. Following the British statistician Karl Pearson, the Gaussian distribution began to be referred to as the normal distribution.

  De Moivre's 1718 book *The Doctrine of Chances* constituted a major contribution to probability theory. It was in his second important book *Miscellanea Analytica* (*Analytical Miscellany*) of 1730 that he introduced the now called Gaussian distribution.

  We may note that Stirling's formula for $n!$, incorrectly attributed to the Scottish mathematician James Stirling, was actually introduced by de Moivre. De Moivre was also the first to use complex numbers in trigonometry.
- **Newton, Isaac** - see footnote 2, Chap. 1.
- **Pascal, Blaise**, 1623–1662, French mathematician, physicist and religious philosopher. Pascal was born in Clermont-Ferrand, France, the son of Étienne Pascal, a presiding judge of the tax court at Clermont-Ferrand and a respected mathematician. In 1631, after his mother's death, the family moved to Paris where Étienne Pascal devoted himself to the education of his children. Both Pascal and his younger sister Jacqueline were viewed as child prodigies; his sister in writing and Pascal in mathematics. As a young man, between 1642 to 1644, he conceived and constructed a calculating machine to help his father in tax computations following his appointment in 1639 as intendant (local administrator) at Rouen.

---

[2]Edmond is also spelt Edmund.

He constructed a mercury barometer to test the theories of Galileo and Evange-lista Torricelli (an Italian physicist who invented the barometer), which led him to further studies in hydrodynamics and hydrostatics. A particularly significant discovery in hydrostatics is what is now known as Pascal's principle or Pascal's law (pressure applied to a confined liquid is transmitted undiminished throughout the liquid). Based on this principle, Pascal invented the hydraulic press. His book *Traité du triangle arithmétique* deals with founding principles of probability and beginning ideas on calculus. This work is regarded as a significant early step in the development of the theory of probability.

Pascal came from a religious Catholic background and maintained strong religious beliefs throughout his life and by the end of 1653 turned his attention more to religion than to science. On November 23 1654 Pascal experienced a strange mystical conversation which he believed marked the beginning of a new life. In January 1654, he entered the Port-Royal convent (a Jansenist convent - Jansenism is an austere form of Catholicism). Though he never became a full member, he only wrote at the request of the Port-Royal Jansenists, and never again published under his own name.

- **Poisson, Siméon-Denis**, 1781–1840, French mathematician. Poisson was born in Pithiviers, France. His family wanted him to study medicine, but Poisson's inter-est was in mathematics and in 1798 enrolled at the École Polytechnique in Paris. His teachers were Pierre-Simon Laplace and Joseph-Louis Lagrange, with whom he became lifelong friends. In 1802 he was appointed a Professor at the École Polytechnique, a post he left in 1808 to take up a position as an astronomer at the Bureau of Longitudes. When the Faculty of Sciences was created in 1809 he was offered the position of professor of pure mathematics.

Poisson made many contributions to mathematics, physics and probability. In 1811 he published an authoritative book on mechanics *Traité de mécanique* (*Treatise on Mechanics*), published again in 1833. In electrostatics he extended the work of Laplace, developing what is now called the Poisson equation. He also extended the work of Laplace and Lagrange on the stability of the solar system. In 1831, he published his theory on capillary action *Théorie nouvelle de l'action capillaire* (*A New Theory of Capillary Action*). Later, in 1835, Poisson published his work on heat *Théorie mathématique de la chaleur* (*Mathematical Theory of Heat*). In pure mathematics Pascal wrote important papers on definite integrals and made important advances in Fourier analysis.

Poisson's contribution to the theory of probability is of particular significance. He introduced what is now called the Poisson distribution as an approximation to the binomial distribution which in this role proved to be of great practical utility. In recent times, it has been found that radioactive counts in a given time intervals or counts of cars along a road in a given time interval, are described by the Poisson distribution. He also contributed to the derivation of the law of large numbers. He presented this work in his 1837 book *Recherches sur la probabilité des jugements en matière criminelle et en matière civile* (*Research on the probability of criminal and civil verdicts*) in a social rather than a scientific context.

- **Vernier** See footnote 4, Chap. 1.

Appendix E
Solutions

Chapter 3

1. (a) $\sigma = 0.0283\,\text{mm}$,　(b) $s_m = 0.00756\,\text{mm}$
2. $m = 1.45$,　$(\bar{x}, \bar{y}) = (7.00, 9.48)$,　$c = -0.667$,　$s_m = 0.03$
3. (a) $(150.0 \pm 0.5)\,\text{g}$　(b) $(50.0 \pm 0.5)\,\text{g}$
4. $(350.0 \pm 0.7)\,\text{g}$
5. $(200 \pm 3)\,\text{cm}^3$
6. (a) Relative error $= 0.0687$ (3 sf)　(b) Percentage error $= 6.87\%$ (3 sf)
7. $(670 \pm 46)\,\text{cm}^3$

Chapter 5

1. {numbers 1 to 36} and {red, black}
2. $\frac{3}{52}$
3. By direct counting the probabilities $P(A_1) = \frac{6}{52}$, $P(A_2) = \frac{13}{52}$. Then, clearly, $P(A_1) \geq P(A_2)$. Also $P(A_1 - A_2) = P(A_1) - P(A_2) = \frac{7}{12}$. By direct counting, the number of ways of getting A_2 but not A_1 is 7 giving a probability $P(A_1 - A_2)$ of getting A_2 but not A_1 of $\frac{7}{12}$ confirming the result from Theorem 5.2.1.
4. By counting desired occurrences in each case we find the probabilities to be $P(A) = \frac{4}{52}$, $P(A_1) = \frac{1}{52}$, $P(A_2) = \frac{1}{52}$, $P(A_3) = \frac{1}{52}$ and $P(A_4) = \frac{1}{52}$. Then

$$P(A) = P(A_1) + P(A_2) + P(A_3) + P(A_4) = \frac{1}{52} + \frac{1}{52} + \frac{1}{52} + \frac{1}{52} = \frac{4}{52}$$

in agreement with Theorem 5.2.4.
5. The probability of choosing a club is $P(A) = \frac{13}{52}$, while the probability of drawing a card numbered 5 to 10 is $P(B) = \frac{24}{52}$. The probability of drawing a club numbered 5 to 10 is $P(A \cap B) = \frac{6}{52}$. Then

$$P(A \cup B) = P(A) + P(B) - P(A \cap B) = \frac{13}{52} + \frac{24}{52} - \frac{6}{52} = \frac{31}{52}$$

© Springer Nature Switzerland AG 2018
P. N. Kaloyerou, *Basic Concepts of Data and Error Analysis*,
https://doi.org/10.1007/978-3-319-95876-7

The probability of drawing a club or a card from 5 to 10 by counting desired outcomes is $P(A \cup B) = \frac{31}{52}$, where clubs 5 to 10 are only counted once, confirming the answer obtained by Theorem 5.2.5.

6. The probabilities, by direct counting are $P(A) = \frac{13}{52}$, $P(A \cap B) = \frac{6}{52}$ and $P(A \cap B') = \frac{7}{52}$, since event B' consists of all numbers from 1 to 4 and 11 to 13. Substituting into Theorem 5.2.6 we get

$$P(A) = P(A \cap B) + P(A \cap B') = \frac{6}{52} + \frac{7}{52} = \frac{13}{52},$$

in agreement with Theorem 5.2.6.

7. By counting desired outcomes we get $P(A) = \frac{13}{52}$, $P(A_1 \cap A) = \frac{3}{52}$, $P(A_2 \cap A) = \frac{4}{52}$, $P(A_3 \cap A) = \frac{2}{52}$ and $P(A_4 \cap A) = \frac{4}{52}$, then

$$P(A) = P(A \cap A_1) + P(A \cap A_2) + P(A \cap A_3) = \frac{3}{52} + \frac{4}{52} + \frac{2}{52} + \frac{4}{52} = \frac{13}{52}$$

in agreement with Theorem 5.2.7.

8. $P(B|A) = P(Y|Y) = \frac{4}{7}$

9. $P(A \cap B) = P(B|A)P(A) = \frac{4}{7} \times \frac{4}{7} = \frac{16}{49}$

10. $P(B|A) = \frac{1}{2}$

11. $P(A \cap B) = \frac{2}{7}$

12. With replacement: $P(A \cap B) = \frac{1}{169}$; without replacement: $P(A \cap B) = \frac{4}{663}$.

13. With replacement: $P(A_1 \cap A_2 \cap A_3) = \frac{1}{2197}$; Without replacement: $P(A_1 \cap A_2 \cap A_3) = \frac{1}{5525}$

14. $_nP_r = 665,280$.

15. $_nC_r = 924$

16. $P(2Y, 2G) = \frac{3}{7}$

17. $p(0) = \frac{1}{8}$, $p(1) = \frac{3}{8}$, $p(2) = \frac{3}{8}$ and $p(3) = \frac{1}{8}$.

18.

$$F(x) = \begin{cases} 0 & \text{for } -\infty < x < 0 \\ \frac{1}{8} & \text{for } 0 \leq x < 1 \\ \frac{1}{8} + \frac{3}{8} = \frac{4}{8} & \text{for } 1 \leq x < 2 \\ \frac{1}{8} + \frac{3}{8} + \frac{3}{8} = \frac{7}{8} & \text{for } 2 \leq x < 3 \\ \frac{1}{8} + \frac{3}{8} + \frac{3}{8} + \frac{1}{8} = 1 & \text{for } 3 \leq x < \infty \end{cases}$$

19.

x	1	2	3	4	5	6	8	9	10
$p(x)$	1/36	2/36	2/36	3/36	2/36	4/36	2/36	1/36	2/36
$F(x)$	1/36	3/36	5/36	8/36	10/36	14/36	16/36	17/36	19/36

x	12	15	16	18	20	24	25	30	36
$p(x)$	4/36	2/36	1/36	2/36	2/36	2/36	2/36	2/36	1/36
$F(x)$	23/36	25/36	26/36	28/36	29/36	31/36	33/36	35/36	36/36

20. (a)

$$F(x) = \begin{cases} \int_{-\infty}^{-1} 0 \, dx = 0 \ \text{ for } \ x < -1 \\[2ex] \frac{1}{\sqrt{\pi}\,\mathrm{Erf}(1)} \int_{-1}^{x} e^{-v^2} \, dv = \frac{\mathrm{Erf}(1)+\mathrm{Erf}(x)}{\mathrm{Erf}(1)} \ \text{ for } \ -1 \le x \le 1 \\[2ex] \int_{1}^{\infty} 0 \, dx = 0 \ \text{ for } \ x > 1. \end{cases}$$

(b) $P(-0.5 \le X \le 0.5) = 0.618$,

21. (a)
$p(0,0) = 0, \ p(0,1) = \frac{4}{84}, \ p(0,2) = \frac{12}{84}, \ p(0,3) = \frac{4}{84}, \ p(1,0) = \frac{3}{84},$

$p(1,1) = \frac{24}{84}, \ p(1,2) = \frac{18}{84}, \ p(2,0) = \frac{6}{84}, \ p(2,1) = \frac{12}{84}, \ p(3,0) = \frac{1}{84}.$

$p_x(0) = \frac{20}{84}, \ p_x(1) = \frac{45}{84}, \ p_x(2) = \frac{18}{84}, \ p_x(3) = \frac{1}{84},$

(b)
$p_y(0) = \frac{10}{84}, \ p_y(1) = \frac{40}{84}, \ p_y(2) = \frac{30}{84}, \ p_y(0) = \frac{4}{84}$

(c)

$X \backslash Y$	$y_1 = 0$	$y_2 = 1$	$y_3 = 2$	$y_4 = 3$	$p_x(x_i)$
$x_1 = 0$	$p(0,0) = 0$	$p(0,1) = \frac{4}{84}$	$p(0,2) = \frac{12}{84}$	$p(0,3) = \frac{4}{84}$	$p_x(0) = \frac{20}{84}$
$x_2 = 1$	$p(1,0) = \frac{3}{84}$	$p(1,1) = \frac{24}{84}$	$p(1,2) = \frac{18}{84}$	—	$p_x(1) = \frac{45}{84}$
$x_3 = 2$	$p(2,0) = \frac{6}{84}$	$p(2,1) = \frac{12}{84}$	—	—	$p_x(2) = \frac{18}{84}$
$x_4 = 2$	$p(3,0) = \frac{1}{84}$	—	—	—	$p_x(2) = \frac{1}{84}$
$p(y_j)$	$p_y(0) = \frac{10}{84}$	$p_y(1) = \frac{40}{84}$	$p_y(2) = \frac{30}{84}$	$p_y(3)\frac{4}{84}$	1

22. (a) $P(X \ge 1, Y \le 1) = 0.01914$, (b) $P(X \le Y) = 0.5$,

(c) $P(X \le a) = \frac{1}{2}\left[1 + [\mathrm{Erf}(\sqrt{2}\,a)]\right]$

23. $p(0|1) = \frac{4}{7}, \quad p(1|1) = \frac{3}{7}$

24.
$$p(x|y) = \begin{cases} \frac{(2x-4)(3y-5)}{(15-9y)} & \text{for } 0 \le x \le 1,\ 0 \le y \le 1 \\ \\ 0 & \text{otherwise} \end{cases}$$

25. $\langle X \rangle = 1.5$
26. $\langle d \rangle = 0.571\,\text{mm}$
27. $\langle X \rangle = \frac{3}{2}$
28. $\langle X \rangle = 0.7034\,\text{m}, \langle g(X) \rangle = 0.3507\,\text{m}^3$.
29. $\sigma^2 = 14, \sigma = \sqrt{14}$
30. $\sigma^2 = 0.02667, \sigma = 0.1633$
31. $\mu_X = 0.7619, \mu_Y = 0.8452, \sigma_X^2 = 0.5567, \sigma_Y^2 = 0.7938,$

$\sigma_X = 0.7461, \sigma_Y = 0.8910, \sigma_{XY} = -0.2665, \rho = -0.4009$

32. $\mu_X = 2.1, \mu_Y = 1.8, \sigma_X^2 = 0.45, \sigma_Y^2 = 0.66, \sigma_X = 0.6708, \sigma_Y = 0.8124,$
$\sigma_{XY} = -0.03, \rho = -0.05505.$

33. $P(X = 0) = \frac{1}{8}, \ P(X = 1) = \frac{3}{8}, \ P(X = 2) = \frac{3}{8}, \ P(X = 3) = \frac{1}{8}$

34. 0.99%
35. $PB(0) = 0.3670, \ PP(0) = 0.3679, \ PB(1) = 0.3688, \ PP(1) = 0.3679,$

$PB(2) = 0.1844, \ PP(2) = 0.1839, \ PB(5) = 0.002982, \ PP(5) = 0.003066$
36. $P(\le 2) = 0.07677$
37. For (a) and (b) $P(2 \le X \le 11) = 0.8186$
38. (a) $P(14 \le X \le 17) = 0.05302$, (b) $P(13.5 \le X \le 17.5) = 0.05836$.

Index

© Springer Nature Switzerland AG 2018
P. N. Kaloyerou, *Basic Concepts of Data and Error Analysis*,
https://doi.org/10.1007/978-3-319-95876-7

Printed in the United States
By Bookmasters